Weber's Thermodynamics Notes

Fred J. Weber, P.E.

Book information

Weber's Thermodynamics Notes

Published by:
Wrench Time, Inc.
Orlando, Fl.
http://www.wrenchtime.com

ISBN-13 978-0-578-43694-4

Library of Congress
Copyright Office
Registration Number TX 8-540-918

Background of the Author

The background of any individual is really the sum of the effects that certain people have on their lives. For me, it was:

my wife, Joyce
(you don't want to miss anything in life)
my father, Albert
(he was the best production manager)
my mother, Margaret
(the best family CEO, she met the budget annually)
my brother, William
(a teamster, and knows we still need unions)
my sister, Diane
(computer manager, she knows everything isn't ones and zeros)

Also, the hundreds of people in maintenance and construction that have bailed me out throughout my career by:

- teaching me that everything was not in my engineering books
- solving technical problems while making me look good doing it
- questioning my decision to make the job safer and easier

So, if you don't like me, blame them!

Table of Contents

Welcome to Weber's Thermodynamics notes

As an engineering student back when slide rulers were popular, I struggled to grasp thermodynamics. One reason for the confusion is the terminology used in thermodynamics:

Entropy - the measure of disorder or the unavailable energy. Pretty confusing.

Free energy – the amount of energy one can extract from a system. In thermodynamics you will learn quickly that nothing is free.

Carnot cycle – the most efficient thermodynamic cycle developed by Nicolas Leonard Sadi Carnot while using two constant temperature processes.

So how is this book different than all the other books on thermodynamics?
This book is based on two statements:

"All energy has two components: "work energy" and "heat energy"

"Work can create heat… but heat can not create Work."

So now, what is required to understand this new concept on thermodynamics?
With an engineering background, I might offer a ton of calculus equations with integrals and derivatives to explain it. However, the purpose of this book is to look beyond equations and formulas to develop a common sense, systematic method for someone to follow.

Who are the targets of this book?

Anyone who has been challenged to calculate a thermodynamic problem. It could be an engineer who is reviewing the performance of a process at their plant. Or the reader might be a journeyman mechanic who suddenly finds himself in charge of reviewing the performance of HVAC equipment. In short, this book is designed for any person interested in just finding an easier approach to thermodynamics.

So why did I write this book? Three reasons:

One…to provide a simple format that can be implemented to help you and I grasp the fundamentals of thermodynamics.

Two… to understand energy, the demands and abuses.

Three… to challenge you to think outside of the Carnot cycle box

Does my approach work?

I don't know, because I still have a lot to learn about this subject. This book is simply an accumulation of some of my handwritten notes I put together. Because this book is a new concept on thermodynamics, the information is just one person's view on Thermodynamics.

Therefore, the information in this book has:

No peer review

No test of data to support the information

Not been edited, so there will be errors

And for these reasons the information in this book is to be used:

"for educational use only"

Please confirm the results using industry standards.

Finally, if nothing else the book gets you thinking about thermodynamics.

To quote Eleanor Roosevelt…

"Great minds discuss ideas; average minds discuss events; small minds discuss people."

Acknowledgement:

"The IAPWS Formulation 1995 for the Thermodynamic Properties of Ordinary Water Substance for General and Scientific Use,"

Physical and Chemical Properties Division

National Institute of Standards and Technology

1 Thermodynamics Properties

"Work can create heat... but heat can not create Work."

Fred J. Weber

1.1 Introduction to Thermodynamics

The word thermodynamics can be defined by Webster's dictionary as "the science that deals with the relationship of heat and mechanical energy and the conversion of one another". The Weber definition of thermodynamics is "the science that deals with work and heat energy within a medium and the interaction of heat and work between two mediums".

Regardless of which definition you use, thermodynamics can typically be described why things get hot or become cold. The applications of thermodynamics and the transfer of heat and work can be numerous, which can include boilers, air conditioning system and refrigerator, you get the idea.

The rules for why things get hot and cold are defined by the laws of thermodynamics.

Discussion of the laws typically starts with the definition of three key words from Webster's dictionary:

Energy… "capacity for performing work"

Work… " the transference of energy when a force produces a movement of a body"

Heat… " a form of energy that causes a body to rise in temperature, to fuse , to evaporate or to expand."

For the last century the laws of thermodynamics looked like this:

The 1st law of Thermodynamics is typically defined as the Conservation of Energy which states: "that energy can neither be created nor destroyed but only converted from one form to another"

The 2nd law of Thermodynamics deals with the decay of available energy to an unavailable condition. This law is defined by the statements of three distinguished men of science.

Rudolf Clausius statement…"It is impossible for a self-acting machine unaided by any external agency to convey heat form one body to another at higher temperature"

Lord Kelvin statement…" It is impossible by means of inanimate material agency to derive mechanical effect from any portion of matter by cooling it below the temperature of surrounding objects"

Max Planck statement… "It is impossible to construct an engine which will work in a complete cycle and produce no effect except the raising of a weight and the cooling of a heat reservoir."

2nd law summary….Simply no perpetual machines allowed. "The work effect can't be greater than the energy supplied."

The equation used to describe the 1st law is: $\Delta U = Q - W$

ΔU = change in internal energy

W = $(p\ \Delta V)$ = net work done by the system

Q = net heat added to the system

Typical Block Diagram

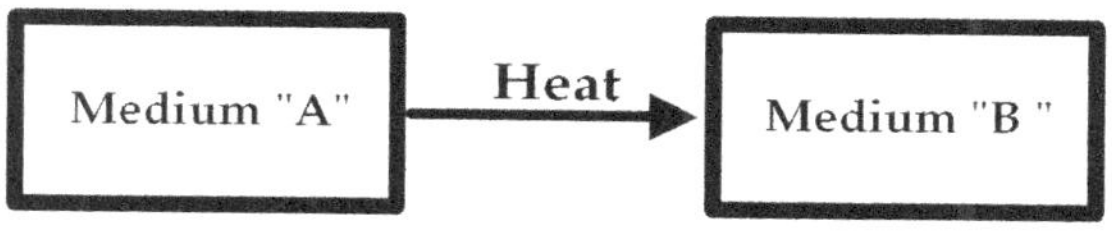

The Wrench Time approach to the Laws of Energy and Thermodynamics is based on one simply concept: All ***Energy*** has two components: ***Work Energy and Heat Energy***

This approach starts by redefining three words…***Energy, Work and Heat*** and adding two other words… ***Work Energy and Heat Energy.***

Energy (E)… capacity available and unavailable for performing work. **(Btu)**

Work (W)… change in work energy ; change in energy available for performing work **(Btu)**

Heat (Q)… change in heat energy ; change in energy unavailable for performing work **(Btu)**

Work Energy (Ew) … capacity of energy available for performing work **(Btu)**

Heat energy (Eq) … capacity of energy unavailable for performing work **(Btu)**

Specific Work Energy (ew) … capacity of energy available for performing work **(Btu / lbm)**

Specific Heat energy (eq) … capacity of energy unavailable for performing work **(Btu / lbm)**

Apply these new terms to develop the Wrench Time approach to Energy and Thermodynamics.

Weber's General Thermodynamics Equations

All energy has two components: work energy and heat energy

Energy = Work Energy + Heat Energy

$E = Ew + Eq$

Change in Energy = change in Work energy + the change in Heat energy

$\Delta E = \Delta Ew + \Delta Eq$

Work is equal to the change in work energy

Work = change in Work Energy

$W = \Delta Ew$

Heat is equal to the change in heat energy

Heat = change in Heat Energy

$Q = \Delta Eq$

Change in Energy = change in Work energy + the change in Heat energy = Work + Heat

$$\Delta E = \Delta Ew + \Delta Eq = W + Q$$

The laws of energy and thermodynamics are typically defined by one medium and one equation. The problem is the 1st and 2nd law of thermodynamics talk about the transfer of energy between two mediums. Therefore, the wrench time approach requires…two mediums, two equations, two sets of data and a different block diagram to show both mediums.

Wrench Time System Diagram

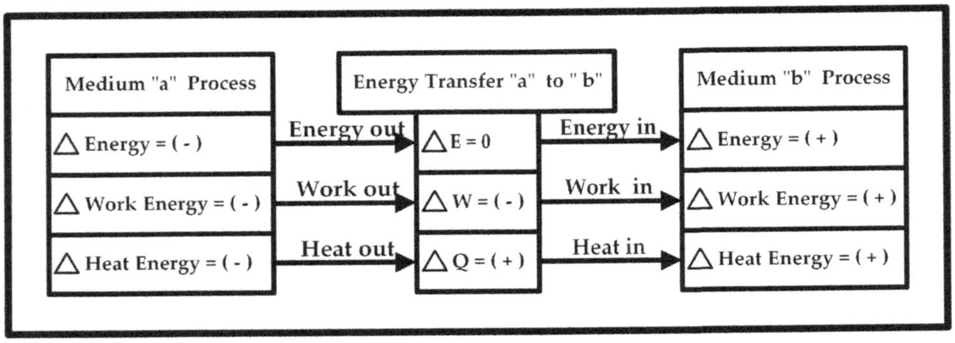

The next tool required for understanding the transfer of energy between two mediums is a common sense sign convention:

(+) = added or in....energy in, energy added, work in, work added, heat in, heat added

(-) = removed or out...energy out, energy removed, work out, work removed, heat removed

(1) = initial condition

(2) = final condition

Weber's 1st Law of Thermodynamics

"Energy can't be created or destroyed". This statement is a copy of **Newton's Conservation of Energy and the First Law of Thermodynamics written by Rudolf Clausius, William Thomson and Lord Kelvin.** Please take the time to read about these scientific legends. This law can be represented by the top portion of the Wrench Time System Diagram

Weber's 1st Law of Thermodynamics

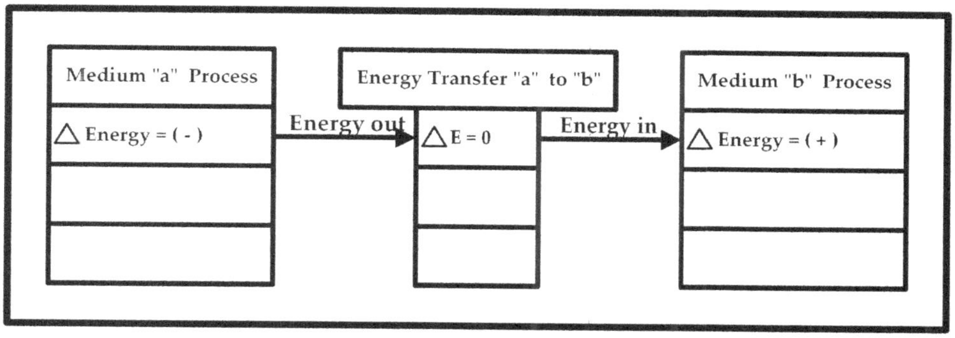

The equation that represents the 1st law:

$$\Delta Ea + \Delta Eb = 0$$ **: energy can't be created or destroyed**

Change in energy medium "a" = Δ Ea

Change in energy medium "b" = Δ Eb

Total change in energy is zero = $\Delta Ea + \Delta Eb = 0$

Energy is removed from medium "a" ($-\Delta$ Ea) and transferred to medium "b" ($+\Delta$ Eb) = 0

Weber's 2nd Law of Thermodynamics

"Work can create heat… but heat can not create Work." This statement represents the energy available to do work transferred out (-) from medium "a" must be greater than or equal to the energy available to do work transferred in (+) to medium "b". This statement can be represented on the bottom portion of the Wrench Time System Diagram below:

Weber's 2nd Law of Thermodynamics

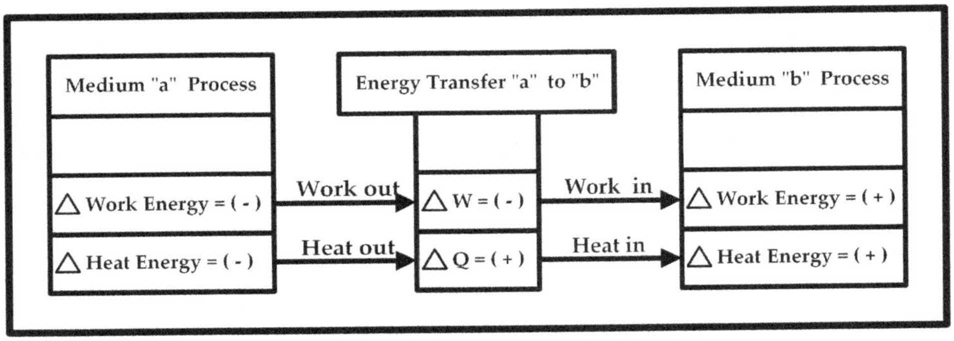

Change in work energy medium "a" = Δ Ewa = Wa

Change in heat energy medium "a" = Δ Eqa = Qa

Change in work energy medium "b" = Δ Ewb = Wb

Change in heat energy medium "b" = Δ Eqb = Qb

Total change in work = (-) = (Δ Ewa + Δ Ewb) = (Wa + Wb)

Total change in heat = (+) = (Δ Eqa + Δ Eqb) = (Qa + Qb)

The second law states that the interact of two mediums "can not" increase the available work of the system then (net work) is equal to or less than 0…(-) therefore (net heat) = 0 or (+).

Weber's 3rd Law of Thermodynamics

Energy moves from a higher level to a lower level, therefore:

The specific energy of medium "a" is greater than or equal to medium "b" (ea > eb)

The specific work energy of medium "a" is greater than medium "b" (ewa > ewb)

While the focus of this book is on thermodynamics, the Weber's Laws of Energy can be applied to all energy such as Thermal, kinetic, potential and etc. :

If ALL energy is involved in this general equation then

Energy = Thermal Energy + Kinetic Energy + Potential Energy + Chemical Energy + Mechanical Energy + Electrical Energy +etc.

E = TE + KE + PE + CE + ME + EE + etc

Change in Energy = change in All energy involved

$$\Delta E = \Delta TE + \Delta KE + \Delta PE + \Delta CE + \Delta ME + \Delta EE$$

If ALL energy has two components: work energy and heat energy then

Work = Change in Work Energy involved

$$W = \Delta TEw + \Delta KEw + \Delta PEw + \Delta CEw + \Delta MEw + \Delta EEw$$

Heat = Change in Heat Energy involved

$$Q = \Delta TEq + \Delta KEq + \Delta PEq + \Delta CEq + \Delta MEq + \Delta EEq$$

Change in Energy = change in Work energy + the change in Heat energy = Work + Heat

$$\Delta E = \Delta Ew + \Delta Eq = W + Q$$

This approach could be used to explain the energy change of a bouncing ball or the effects of friction but for now focus of this book is thermodynamics. Therefore the Weber's Laws of Thermodynamics are focused on the thermal energy and its changes.

Change in Energy equal to change in thermal energy…$\Delta E = \Delta TE$

Change in Work Energy equal to change in thermal work energy … $\Delta Ew = \Delta TEw$

Change in Heat Energy equal to change in thermal heat energy … $\Delta Eq = \Delta TEq$

The following example below of the interaction between two containers of water can be used to get an understanding of the Weber's Law of Thermodynamics

Medium "a" … mass 2 lbm @ 580 (°R). (120 °F)

Medium "b" …mass of 2 lbm @ 540 (°R). (80 °F)

Medium "A"	Temp	Energy	Heat Energy	Work Energy
1-Initial	580	88.4	81.3	7.1
2-Final	560	68.4	64.1	4.3
(2 - 1)	- 20.0	- 20.0	- 17.2	- 2.8
2 lbm /hr		Δ Eb = -40	Qb= -34.4	Wb= -5.6

Net change	Δ E = 0	Δ Q = +1.4	ΔW = -1.4

Medium "B"	Temp	Energy	Heat Energy	Work Energy
1-Initial	540	48.5	46.2	2.3
2-Final	560	68.4	64.1	4.4
(2 - 1)	20.0	20.0	17.9	2.1
2 lbm		Δ Eb = +40	Qb= +35.8	Wb= +4.2

Spontaneous interaction between media "a" and "b":

1. **Energy moves from a higher level "a" to a lower level "b"**
   specific energy "a"...88.4 > "b"...48.5
   specific work energy "a"...7.1 > "b"...2.3
2. **Total energy change equal to zero** = - 40 + 40 = 0
3. **Total work and heat** = - 5.6+ (- 34.4) + 35.8 + 4.2 = 0
4. **(-) net work out** = -5.6 + 4.2 = - 1.4
5. **(+) net heat in** = -34.4 + 35.8 = + 1.4
6. **Net work + Net heat** = (-1.4) + (+1.4) =0

Before we go into detail discussion of Energy, Work and Heat we need to do a little Thermodynamics housekeeping.

1. Define terms and units used throughout the book.
2. Thermo-diagram, a graphic representation of the properties and data.
3. Graphs and Data recommended when solving thermodynamic problems.
4. Discussion of the terms property, state, process, cycle and system

1.2 Definitions of thermodynamics

The following is a brief list of terms used though out the book.

Cycles... a series of processes of a medium forming a closed loop. One requirement for a cycle, the medium returns to its initial state. Properties return to (T1, P1, S1, H1 and mass flow)

Density (ρ ... lbm / ft^3) the property that defines the mass per unit volume

Enthalpy (**H**...Btu) the total energy of a medium H= U + (0.185)PV at defined by the sum of the (mean temperature times entropy) along a constant pressure line. The total energy of a medium available and unavailable to work (hw + hq) **specific enthalpy** (**h**... Btu / lbm)

Enthalpy heat.... enthalpy in a medium unavailable to do work.

(**HQ...** enthalpy heat (Btu) and **hq...** specific enthalpy heat (Btu / lbm)

Enthalpy work...enthalpy in the medium available to do work.

(**HW...** enthalpy work (Btu) and **hw...** specific enthalpy work (Btu / lbm))

Entropy (**S**... Btu/ °R) ... property that relates temperature to enthalpy and internal energy

Heat (Q... Btu) the energy transferred from a medium unavailable to do work.

(**Qin...** heat added to a medium. **Qout...** heat removed from a medium)

Heat capacity... the linear change in enthalpy divided by the change in temperature (Btu/°R)

Internal Energy (**U...**Btu) ...kinetic energy of molecules of a medium

Internal heat energy ... internal energy in the medium unavailable to do work.

(**UQ...** internal heat energy (Btu) and **uq...** specific Internal heat energy (Btu / lbm))

Internal work energy .... Internal energy in the medium available to do work.

(**UW...** internal work energy (Btu) and **uw...** specific Internal work energy (Btu / lbm))

Isentropic... a process the medium remains at a constant specific entropy (s1 = s2) (Btu/lbm-°R)

Isobaric... a process the medium remains at a constant pressure (P1 = P2) (psia)

Isothermal...... a process the medium remains at a constant temperature (T1 = T2) (°R)

Isovolumic ... a process the medium remains at a constant specific volume (v1 = v2) (ft3/lbm)

Pressure (**P**... lbf / in^2 absolute) (psia)... is the force per unit area from a medium.

Process. . the change of a medium's property value at a particular time

Property... a particular or calculated attribute of a medium.

a. **Extensive property**...is a property of a medium effective by the mass. (Enthalpy... h (Btu))

b..**Intensive property**...is a property of a medium **<u>not</u>** effective by the mass (Temperature)

State... properties and their values of a medium at a particular time

System.... the energy interaction between two different mediums. The transfer of energy, work and heat between medium.

Temperature... property related to the kinetic energy inside the medium.

Work (W... Btu) the energy transferred from a medium available to do work.

(**Win...** work added to a medium. **Wout...** work removed from a medium)

<u>Misc. Definitions</u>

Units

Throughout this book we will be using the English units to keep the focus on thermodynamics instead of spending time doing unit conversions. (Examples...psia, °R, (ft^3/lbm and Btu)

Sign Convention

(+) = added or in....energy in, energy added, work in, work added, heat in, heat added

(-) = removed or out...energy out, energy removed, work out, work removed, heat removed

(1) = initial condition

(2) = final condition

Note: This common sense convention is different than most thermodynamics books.

1.3 Thermo-diagram

The thermo-diagram is a graphic snap shot of a thermodynamic process of a medium. The purpose of the thermo-diagram is to show the thermodynamic properties and their value of a medium as the process changes. Examples of this could be the inlet and outlet property values of a heat exchanger, compressor or a turbine.

A thermo-diagram can take any shape but should display the following:

1. Properties and units of interest (including calculated properties)
2. Defines the medium
3. Initial values of the medium's properties (including assumed values)
4. Final values of the medium's properties (including assumed values)
5. Change in the values of the medium's properties (Final value – initial values)
6. Define the process
7. Mass or mass flow (value and units)
8. The energy or energy flow entering or leaving the medium due to the process

Here is what a thermo-diagram could look like:

H2O	Temp	Pressure	Volume	Entropy	Enthalpy	uq / hq	hw	Pv work	Tmh
491.69	°R	psia	ft^3/lbm	Btu/lbm-°R	Btu/lbm	Btu/lbm	Btu/lbm	Btu/lbm	°R
1- Initial	1300	100	7.691	1.8626	1451.2	915.8	535.4	142.3	779.13
2- Final	900	100	5.203	1.6772	1249.7	824.7	425.0	96.3	745.11
(2 - 1)	-400.0	0.0	-2.49	-0.1854	-201.5	-91.2	-110.3	-46.0	-34.0
1	lbm /hr		-2.49	-0.1854	-201.5	-91.2	-110.3	-46.0	

The function of a thermo-diagram is to display:

1. Properties and units of interest (including calculated properties)

The properties used on a thermo-diagram are the user choice but I do recommend you have the standard properties Temperature, Pressure, Volume, Entropy and Enthalpy. Calculated properties can be added or removed from the thermo-diagram depending your application. For now we will use uq/hq, hw, Pv, and Tmh and define them later in the book. Also notice the units for each property is listed. In this example we are using English units (psia, Btu/lbm).

2. Defines the medium

Know the medium. This example it is **H2O** (water/steam). Other medium could be ammonia, 134 or propane.

3. Initial values of the medium's properties (including assumed values)

These are the initial values of the medium's properties shown along the row labeled **"1-Initial"**. In our example we did not assume any property value but if would have we need to highlight, circle or by changing the text font. Use any method, just so the person looking at the data knows that the property value was assumed.

4. Final values of the medium's properties (including assumed values)

These are the final values of the medium's properties shown along the row labeled **"2-Final"**. Remember if the data for a particular property value was assumed let the observer know it.

5. Change in the values of the medium's properties (Final value – initial values)

These values represent the change in property value from the initial ("1-Initial") and final ("2-final") state along the row labeled "**(2-1)**" .

6. Defines the process.

What process the medium is goes through is important. The process is defined by which property has no change between the initial and final state…"**(2-1)**". In our example the property with zero change is pressure therefore the process is at a constant pressure (isobaric). If another property changes were zero the process would be:

A. Zero temperature change would be constant temperature (isothermal)

B. Zero pressure change would be constant pressure (isobaric)

C. Zero volume change would be constant temperature (isometric)

D. Zero entropy change would be constant entropy (isentropic)

E. Zero enthalpy change would be constant enthalpy (throttling)

7. Mass or mass flow (value and units)

This value is either the mass or flow of the medium being changed in the process. In our example we had 5 lbm/hr. The value for mass needs to match the units of the specific property. Enthalpy units are xxx/lbm then mass or mass flow must be lbm/yyyy.

8. The energy or energy flow entering or leaving the medium due to the process

This row of data is the results from (5 lbm/hr) times "**(2-1)**" the change in property value. Example is the change in enthalpy is -112.7 Btu/lbm times the flow rate of 5 lbm/hr…gives a result of 563.5 Btu/hr. Note "lbm" cancels.

Note:

Typically we would subtract 1-2 on most spread sheets but this is not a spreadsheet it is a thermo-diagram. The reasoning for this configuration is:

A. Stacking data for analyzing a cycle or system. If we placed one thermo-diagram above a second on the data at "**2-final**" of the first process becomes the data "**1-Initial**" for the second process which allows the user to stack the information used in analyzing a system (multi-process)

B. Using the results of "**(2-1)**" instead of "(1-2)" we can quickly see if energy is being added or removed from the medium. In our example we see a negative enthalpy change which represents energy is removed from the process. If the results were positive then energy is added to the process.

C. Easy to define the process by looking for zero in the results "**(2-1)**" .

D. User can easily add or remove properties. The properties and calculations selected are user's choice. If the information for the internal energy and density is important, then add it.

E. The ability today to access thermodynamic properties of a medium in a spreadsheet format allows the user to compare and calculate any type of thermodynamic information

While this is one approach to the thermo-diagram, the focus is to supply a snap shot of the medium, its properties, the process and the flow of energy entering or leaving the medium. Below are calculations for a thermo-diagram.

Medium's property

	Temp	Pressure	Volume	Entropy	Enthalpy
	°R	psia	ft³/lbm	Btu/lbm-°R	Btu/lbm
1- Initial	T1	P1	v1	s1	h1
2- Final	T2	P2	v2	s2	h2
(2 - 1)	= T2 - T1	= P2 - P1	= v2 - v1	= s2 - s1	= h2 - h1

Medium's calculated property

	uq / hq	hw	pv	Tmh	shw	shq
	Btu/lbm	Btu/lbm	Btu/lbm	°R	Btu/lbm-°R	Btu/lbm-°R
1- Initial	hq1 = Tr x s1	hw1= h1-hq1	(Pv)1 = 0.185 x P1 x v1	Tmh1= h1 / s1	shw1 = hw1 / s1	shq1 = hq1 / s1
2- Final	hq2 = Tr x s2	hw2 = h2-hq2	(Pv)2 = 0.185 x P2 x v2	Tmh2 = h2 / s2	shw2 = hw2 / s2	shq2 = hq2 / s2
(2 - 1)	= hq2 - hq1	= hw2 - hw1	= (Pv)2 - (Pv)1	= Tmh2 - Tmh1	= shw2 - shw1	= shq2 - shq1

	Int Energy	uq / hq	uw	Tmu	suw	suq
	Btu/lbm	Btu/lbm	Btu/lbm	°R	Btu/lbm-°R	Btu/lbm-°R
1- Initial	u1 = h1 - (Pv)1	uq1 =Tr x s1	uw1 = u1 - uq1	Tmu1 = u1 / s1	suw1 = uw1 / s1	suq1 = uq1 / s1
2- Final	u2 = h2 - (Pv)2	uq2 = Tr x s2	uw2 = u2 - uq2	Tmu2 = u2 / s2	suw2 = uw2 / s2	suq2 = uq2 / s2
(2 - 1)	= u2 - u1	= uq2 - uq1	= uw2 - uw1	= Tmu2 - Tmu1	= suw2 - suw1	= suq2 - suq1

1.3 Property, process, cycle and system

Property... a particular or calculated attribute of a medium.

Term	Property	Units	Specific Property	Units
Temperature	T	(°R)	n/a	n/a
Pressure	P	(psia)	n/a	n/a
Volume	V	(ft^3)	v	(ft^3/ lbm)
Entropy	S	(Btu/ °R)	s	(Btu/ lbm-°R)
Enthalpy	H	(Btu)	h	(Btu/lbm)
Enthalpy Work	Hw	(Btu)	hw	(Btu/lbm)
Enthalpy Heat	Hq	(Btu)	hq	(Btu/lbm)
Internal Energy Heat	Uq	(Btu)	uq	(Btu/lbm)
Internal Energy	U	(Btu)	u	(Btu/lbm)
Internal Energy Work	Uw	(Btu)	uw	(Btu/lbm)
Internal Energy Heat	Uq	(Btu)	uq	(Btu/lbm)
PV work	(0.185)PV	Btu	(0.185)Pv	(Btu/lbm)

Energy Transfer Properties

Notice that Temperature and Pressure are intensive properties of the medium. These properties have no specific property associated to it. The other properties have a specific property, the term "lbm" and are effective by the mass. These properties are the energy properties that affect the state of a medium and describe what type of energy that is entering or leaving a medium.

Process... a change in the a property value at a particular time

Snap shot of gauge and instrument readings within a particular time frame (hour) across a device such as a heat exchanger or turbine.

H2O	Temp	Pressure	Volume	Entropy	Enthalpy	uq / hq	hw	Pv work	Tmh
491.69	°R	psia	ft^3/lbm	Btu/lbm-°R	Btu/lbm	Btu/lbm	Btu/lbm	Btu/lbm	°R
1- Initial	1300	100	7.691	1.8626	1451.2	915.8	535.4	142.3	779.13
2- Final	900	100	5.203	1.6772	1249.7	824.7	425.0	96.3	745.11
(2 - 1)	-400.0	0.0	-2.49	-0.1854	-201.5	-91.2	-110.3	-46.0	-34.0
1	lbm /hr		-2.49	-0.1854	-201.5	-91.2	-110.3	-46.0	

Record the medium's properties before and after a process to understand the energy changes of the medium. By doing this we are to define the heat and work energy of a medium such as enthalpy heat and enthalpy work later in our discussion. Examples of some processes in industry today are heat exchanger, AC compressor or a steam turbine.

Cycle… multi-processes of the same medium

A cycle is a loop of processes interconnected together. The medium starts a state "X" goes through series of processes, transferring heat and work but return to the initial state of "X". Below is an example a cycle of a medium.

Rankine Cycle

Pump

H2O	Temp	Pressure	Volume	Entropy	Enthalpy	uq / hq	hw	Pv work	Tmh
491.69	°R	psia	ft^3/lbm	Btu/lbm-°R	Btu/lbm	Btu/lbm	Btu/lbm	Btu/lbm	°R
1- Initial	585.7	2	0.016	0.1751	94.09	86.10	7.99	0.01	537.29
2- Final	1121.8	2400	0.028	0.9033	719.15	444.15	275.00	12.38	796.12
(2 - 1)	536.1	2398.0	0.01	0.7282	625.07	358.05	267.01	12.38	258.8
1	lbm /hr		0.012	0.7282	625.07	358.05	267.01	12.38	

Boiler

H2O	Temp	Pressure	Volume	Entropy	Enthalpy	uq / hq	hw	Pv work	Tmh
491.69	°R	psia	ft^3/lbm	Btu/lbm-°R	Btu/lbm	Btu/lbm	Btu/lbm	Btu/lbm	°R
1- Initial	1121.8	2400	0.028	0.9033	719.15	444.15	275.00	12.38	796.12
2- Final	2716.2	2400	0.674	1.9207	2243.60	944.39	1299.21	299.31	1168.12
(2 - 1)	1594.4	0.0	0.65	1.0174	1524.45	500.24	1024.21	286.93	372.0
1	lbm /hr		0.646	1.0174	1524.45	500.24	1024.21	286.93	

Turbine

H2O	Temp	Pressure	Volume	Entropy	Enthalpy	uq / hq	hw	Pv work	Tmh
491.69	°R	psia	ft^3/lbm	Btu/lbm-°R	Btu/lbm	Btu/lbm	Btu/lbm	Btu/lbm	°R
1- Initial	2716.2	2400	0.674	1.9207	2243.60	944.39	1299.21	299.31	1168.12
2- Final	585.7	2	173.710	1.9207	1116.50	944.39	172.11	64.27	581.30
(2 - 1)	-2130.5	-2398.0	173.04	0.0000	-1127.10	0.00	-1127.10	-235.04	-586.8
1	lbm /hr		173.036	0.0000	-1127.10	0.00	-1127.10	-235.04	

Condenser

H2O	Temp	Pressure	Volume	Entropy	Enthalpy	uq / hq	hw	Pv work	Tmh
491.69	°R	psia	ft^3/lbm	Btu/lbm-°R	Btu/lbm	Btu/lbm	Btu/lbm	Btu/lbm	°R
1- Initial	585.7	2	173.710	1.9207	1116.50	944.39	172.11	64.27	581.30
2- Final	585.7	2	0.016	0.1751	94.09	86.10	7.99	0.01	537.29
(2 - 1)	0.0	0.0	-173.69	-1.7456	-1022.42	-858.29	-164.13	-64.27	-44.0
1	lbm /hr		-173.694	-1.7456	-1022.42	-858.29	-164.13	-64.27	

The cycle which is the multi-process that interfaces between processes within the same medium. Example of a cycle is the feed water / steam cycle of a power plant. The water / steam cycle:

1. feed water pump
2. Boiler
3. Turbine
4. Condenser
5. Back to feed water pump

<u>System… an interface between two mediums</u>

A system is defines how work and heat energy is transferred between two mediums. Examples of energy transfer:

1. HVAC evaporator to chilled water line
2. Steam condenser to circulating water
3. High pressure steam to turbine

Or simply the energy transfer from a hot glass of water in contact with a cold glass of water.

H2O	Temp	Pressure	Volume	Entropy	Enthalpy	uq / hq	hw	Pv work	Tmh
491.69	°R	psia	ft³/lbm	Btu/lbm-°R	Btu/lbm	Btu/lbm	Btu/lbm	Btu/lbm	°R
1- Initial	580	14.7	0.016	0.16533	88.425	81.29	7.13	0.04	534.84
2- Final	560	14.7	0.016	0.13027	68.445	64.05	4.39	0.04	525.41
(2 - 1)	-20.0	0.0	0.00	-0.03506	-19.980	-17.24	-2.74	0.00	-9.4
2	lbm /hr		0.000	-0.07012	-39.960	-34.48	-5.48	0.00	

Energy	Heat	Work
0.000	1.245	-1.245

H2O	Temp	Pressure	Volume	Entropy	Enthalpy	uq / hq	hw	Pv work	Tmh
491.69	°R	psia	ft³/lbm	Btu/lbm-°R	Btu/lbm	Btu/lbm	Btu/lbm	Btu/lbm	°R
1- Initial	540	14.7	0.016	0.09394	48.465	46.19	2.27	0.04	515.89
2- Final	560	14.7	0.016	0.13027	68.445	64.05	4.39	0.04	525.41
(2 - 1)	20.0	0.0	0.00	0.03633	19.980	17.86	2.12	0.00	9.5
2	lbm /hr		0.000	0.07265	39.960	35.72	4.24	0.00	

1.4 Data and Diagrams

What information, data and diagrams are required to analyze a thermodynamics problem?

The tools required for a Wrench Time approach to thermodynamics. As you analyze any problem in thermodynamics you need to ask yourself:

1. What units of measurement to be used? (Metric or English) Be consistent, select only one. Temperature readings need to be in Rankine or Kelvin.

2. What is the sign convention? Establish a sign convention, though out the book it is:

(+) = added or in....energy in, energy added, work in, work added, heat in, heat added

(-) = removed or out...energy out, energy removed, work out, work removed, heat removed

(1) = initial condition

(2) = final condition

Note: This convention is different than most thermodynamics books.

3. What design of a thermo-diagram is being used to display the data?

4. What is the medium being analyzed? Air, water or R134

5. What is the mass or mass flow rate medium? Be aware of units (Metric or English)

6. What process is the medium going though? Constant pressure or entropy

7. What are the initial properties the medium? T_1, P_1, or H_1

8. What is the final properties and energy the medium? T_2, P_2, or H_2

9. What energy is transferred into and out of the medium? Qin, Qout, Win and Wout

10. What data is unspecified / assumed?

What data is required?

Most processes in industry operate at a constant pressure or between two pressures. Define a temperature range…locate Reference temperature (example 491.69 R) to maximum temperature to be analyzed. Obtain constant pressure data of the medium such as 2, 14.7, 100, 800 and 2400 psia. Below is an example of constant pressure reading of water / steam for 14.7psia.

Top five properties and their units:
Temperature
Pressure
Volume
Entropy
Enthalpy

Display constant pressure data… Most devices operate at a constant pressure or between two pressures

Data starts at Reference Temperature

Display the latent heat data (liquid to vapor)

Eliminate superheated tables: Any temperature reading above the latent heat temperature is superheat.

Temp	Pressure	Volume	Entropy	Enthalpy
°R	psia	ft³/lbm	Btu/lbm-°R	Btu/lbm
491.69	14.7	0.02	0.0000	0.05
500	14.7	0.02	0.0169	8.41
600	14.7	0.02	0.1992	108.42
671.64	14.7	0.02	0.3124	180.30
671.64	14.7	26.80	1.7577	1151.00
700	14.7	28.01	1.7780	1165.00
800	14.7	32.20	1.8417	1212.60
900	14.7	36.32	1.8973	1259.90
1000	14.7	40.41	1.9475	1307.50
1100	14.7	44.49	1.9935	1355.70
1200	14.7	48.56	2.0361	1404.70
1300	14.7	52.63	2.0759	1454.50
1400	14.7	56.69	2.1135	1505.10
1500	14.7	60.75	2.1490	1556.70

Add calculated values that help the end users

Temp	Pressure	Volume	Entropy	Enthalpy	uq / hq	hw	pv	Tmh
°R	psia	ft³/lbm	Btu/lbm-°R	Btu/lbm	Btu/lbm	Btu/lbm	Btu/lbm	°R
491.69	2400	0.016	0.0002	7.2	0.1	7.1	7.1	491.69
500	2400	0.016	0.0167	15.4	8.2	7.2	7.1	919.36
700	2400	0.017	0.3508	214.0	172.5	41.5	7.5	610.13
900	2400	0.019	0.6107	421.2	300.3	120.9	8.4	689.71
1100	2400	0.025	0.8638	675.3	424.7	250.5	11.3	781.71
1121.8	2400	0.028	0.9033	719.2	444.2	275.0	12.4	796.12
1121.8	2400	0.141	1.2452	1102.6	612.3	490.3	62.5	885.48
1300	2400	0.261	1.4490	1345.0	712.5	632.5	115.9	928.23
1500	2400	0.336	1.5526	1489.4	763.4	726.0	149.0	959.29

Below are the latent properties of the medium.

Top five properties and their units:
Temperature, Pressure, Volume, Entropy, and Enthalpy

Display the latent heat data (liquid and vapor)

Data starts at Reference Temperature

Display nominal pressure data... Most devices operate at a constant pressure or between two pressures

End data at critical temperature and pressure

Temperature (°R)	Pressure (psia)	Liquid Volume (ft³/lbm)	Liquid Entropy (Btu/lbm-°R)	Liquid Enthalpy (Btu/lbm)	Vapor Volume (ft³/lbm)	Vapor Entropy (Btu/lbm-°R)	Vapor Enthalpy (Btu/lbm)
491.69	0.1	0.016	0.0000	0.0	3299.400	0.0000	0.0
621.9	5.0	0.016	0.2350	130.3	73.521	1.8451	1131.5
671.6	14.7	0.017	0.3124	180.3	26.795	1.7577	1151.0
787.5	100.0	0.018	0.4746	298.7	4.433	1.6043	1188.3
841.5	200.0	0.018	0.5442	355.7	2.288	1.5470	1199.6
877.0	300.0	0.019	0.5886	394.2	1.544	1.5121	1204.1
904.3	400.0	0.019	0.6221	424.4	1.162	1.4862	1205.8
926.7	500.0	0.020	0.6495	449.8	0.928	1.4652	1205.8
945.9	600.0	0.020	0.6728	472.1	0.770	1.4473	1204.6
962.8	700.0	0.021	0.6933	492.0	0.656	1.4315	1202.7
977.9	800.0	0.021	0.7117	510.2	0.569	1.4171	1200.1
991.7	900.0	0.021	0.7285	527.2	0.501	1.4039	1197.0
1004.3	1000.0	0.022	0.7440	543.0	0.446	1.3916	1193.4
1026.9	1200.0	0.022	0.7721	572.3	0.362	1.3687	1185.0
1046.8	1400.0	0.023	0.7972	599.3	0.302	1.3474	1175.2
1064.6	1600.0	0.024	0.8204	624.6	0.255	1.3271	1164.0
1080.7	1800.0	0.025	0.8421	648.8	0.218	1.3072	1151.4
1095.5	2000.0	0.026	0.8628	672.3	0.188	1.2872	1137.2
1121.8	2400.0	0.028	0.9033	719.2	0.141	1.2452	1102.6
1144.7	2800.0	0.031	0.9470	770.9	0.103	1.1947	1054.4
1164.8	3200.1	0.050	1.0543	897.9	0.050	1.0543	897.9

Values in between liquid and vapor

Quality = "x" = "0" @ liquid and "1" @ vapor

$$h = h_{liquid} + (x)(h_{vapor} - h_{liquid})$$

$$s = s_{liquid} + (x)(s_{vapor} - s_{liquid})$$

$$v = v_{liquid} + (x)(v_{vapor} - v_{liquid})$$

What diagrams of each medium are required?

1. Temperature – specific entropy diagram (don't analyze a thermodynamics problem without it)

2. Temperature – specific enthalpy diagram with hq data

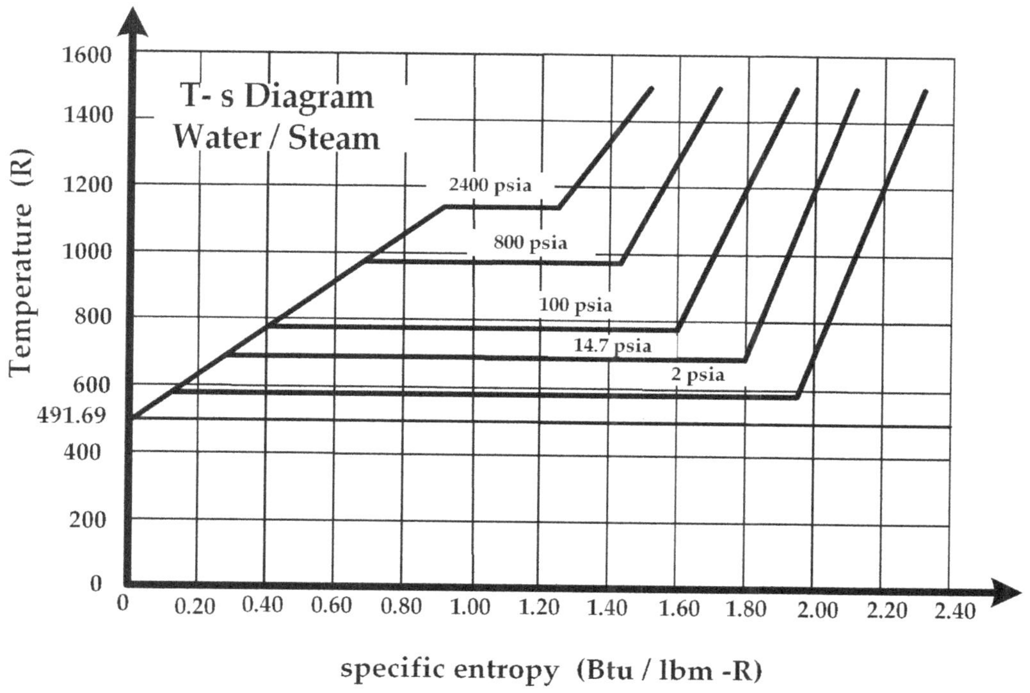

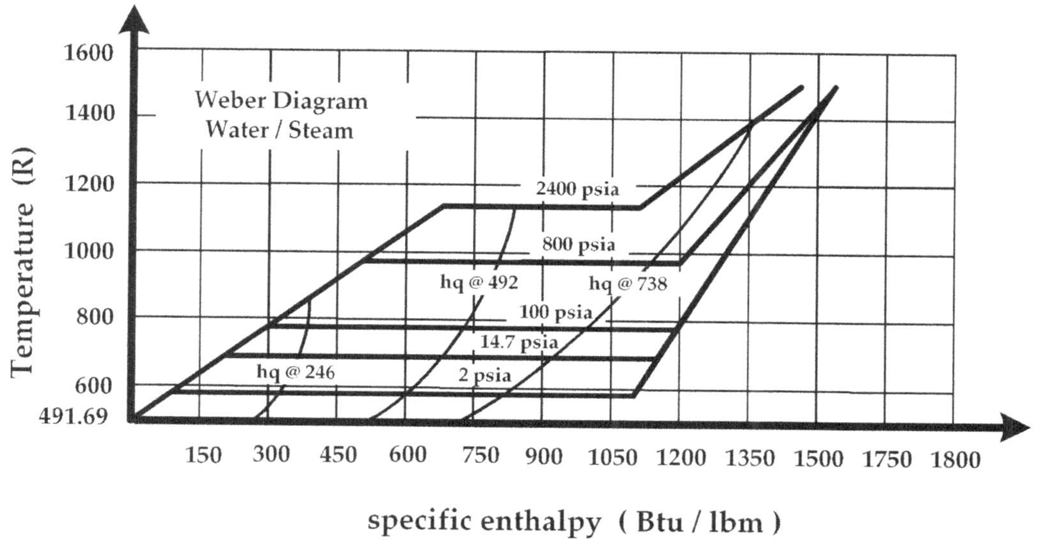

Note: The Weber diagram is a T - h diagram with constant pressure and constant enthalpy heat. Use diagram to locate T , h, p, and hq (s = hq / 491.69)

2 Heat and Work inside the medium

"Energy has two components work energy and heat energy "
Fred J. Weber

2.1 Entropy

The purpose of this module is to discuss entropy, heat energy and work energy inside of a medium. So what is entropy? On a typical water/steam table it may look like this:

Water / Steam @ 14.7 psia

Temp	Pressure	Volume	Entropy	Int. Energy	Enthalpy
°R	psia	ft³/lbm	Btu/lbm-°R	Btu/lbm	Btu/lbm
671.64	14.7	26.795	1.7577	1078.1	1151

Or maybe we need to rephrase the question "What is entropy?" to "How is the value 1.7577 (Btu / lbm-R) calculated ?"

Before we go any farther with the discussion of entropy we need to understand a little math. Assume we have a collected the data from two cars traveling along two different paths. The data and graph below shows the speed and distance recorded for each car. The data for car #1 is contained in curve #1 and the data for car #2 is record on curve #2.

Curve #1

Point	Speed (miles/hr)	Distance miles	??? (???)
0	4	0	0
0 to 1a	10	28	?
1a to 1b	10	120	?
Total	??	148 miles	??

Curve # 2

Point	Speed (miles/hr)	Distance miles	??? (???)
0	4	0	0
0 to2a	8	36	?
2a to 2b	10	90	?
Total	??	126 miles	??

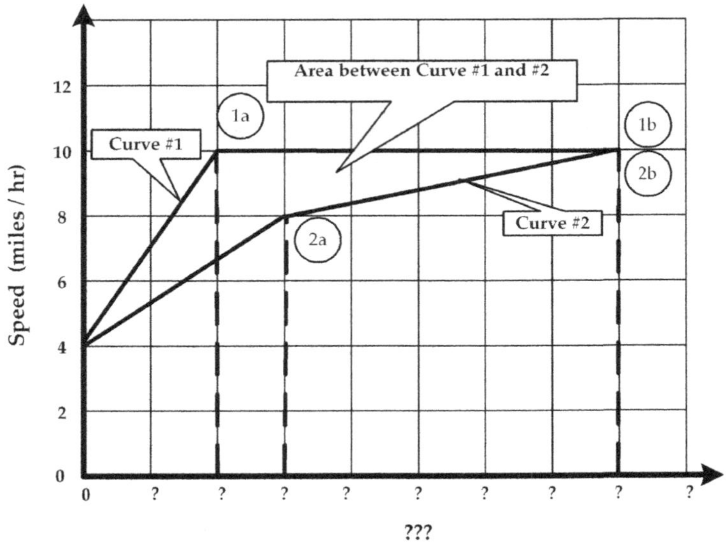

The questions:

1. What property relates distance to speed? ***Time***

Using the equation Distance (miles) = Speed (miles / hours) x Time (hours)

Time (hours) = Distance (miles) / Speed (miles/hr)

2. What is the ***Time*** required each car to travel along the two different paths?

One approach to calculating the ***Time*** both cars traveled is to plot the data on a graph. To calculate the ***Time*** we need to determine the "area under the curve" of each path.

Let's calculate the time required by Car 1 (curve #1) to travel 28 miles with speed change between 4 to 10 miles/hour.

First sketch out the graph using:

Speed (miles / hour) as the y-axis
Time (hours) as the x-axis

Distance (miles) as the area under the curve

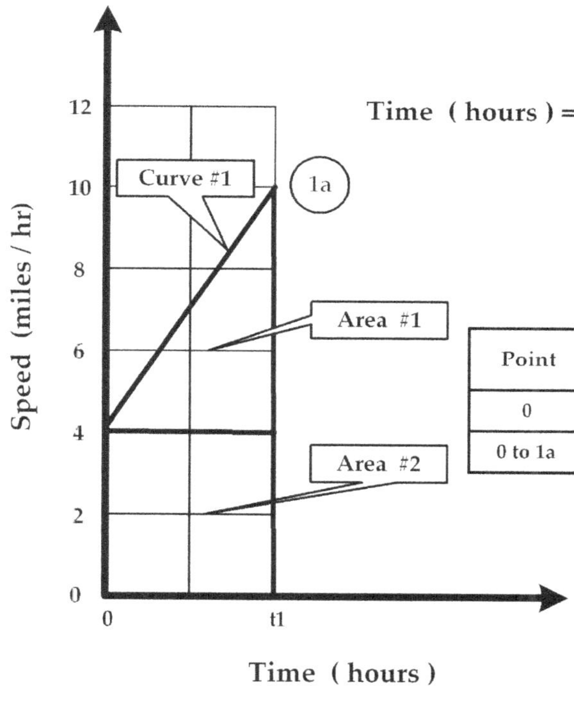

$$\text{Time (hours)} = \frac{\text{Distance (miles)}}{\text{Speed (miles / hours)}}$$

Curve #1

Point	Speed (miles/hr)	Distance miles	Time (hour)
0	4	0	0
0 to 1a	10	28	?

The area under curve can be calculated two ways:

The 1st Approach...using the area of shapes

Total Area = Triangle area (#1) + Rectangle (#2)

$$\Delta \text{ Distance} = (1/2)\,(\text{Speed 1a} - \text{Speed 0})\,(\Delta \text{ Time}) + (\text{Speed 0})(\Delta \text{ Time})$$

$$\Delta \text{ Time} = \frac{\Delta \text{ Distance}}{(1/2)\,(\text{Speed 1a} + \text{Speed 0})}$$

$$\Delta \text{ Time} = \frac{28}{(1/2)\,(10 + 4)} = 4 \text{ (hours)}$$

The 2nd Approach...using the mean value

Total Area = (Ymean)(Δ X) **(Ymean) = (1/2) (y1 + y0)**

$$\Delta \text{ Distance} = (1/2)\,(\text{Speed 1a} + \text{Speed 0})\,(\Delta \text{ Time})$$

$$\Delta \text{ Time} = \frac{\Delta \text{ Distance}}{(1/2)\,(\text{Speed 1a} + \text{Speed 0})} = \frac{28}{(1/2)\,(10 + 4)} = 4 \text{ (hours)}$$

Notice the same time with either approach: 4 hours. Using the mean value calculate the time for the different interval of both cars:

<u>Car 1 (Curve #1):</u>

Time (0 to 1a) = 28 / 7 = 4 hours
Time (1a to 1b) = 120 / 10 = 12 hours
Total time… 4 + 12 =16 hours

<u>Car 2 (Curve #2):</u>

Time (0 to 2a) = 36 / 6 = 6 hours
Time (2a to 2b) = 90 / 9 = 10 hours
Total time…6 + 10 =16 hours

Enter the calculated time values into the charts and modify the graph by replacing "???" with time (hours) values on the x-axis.

Curve #1

Point	Speed (miles/hr)	Distance miles	Time (hours)
0	4	0	0
0 to 1a	10	28	4
1a to 1b	10	120	12
Total	9.25 avg	148 miles	16 hours

Curve # 2

Point	Speed (miles/hr)	Distance miles	Time (hours)
0	4	0	0
0 to2a	8	36	6
2a to 2b	10	90	10
Total	7.88 avg	126 miles	16 hours

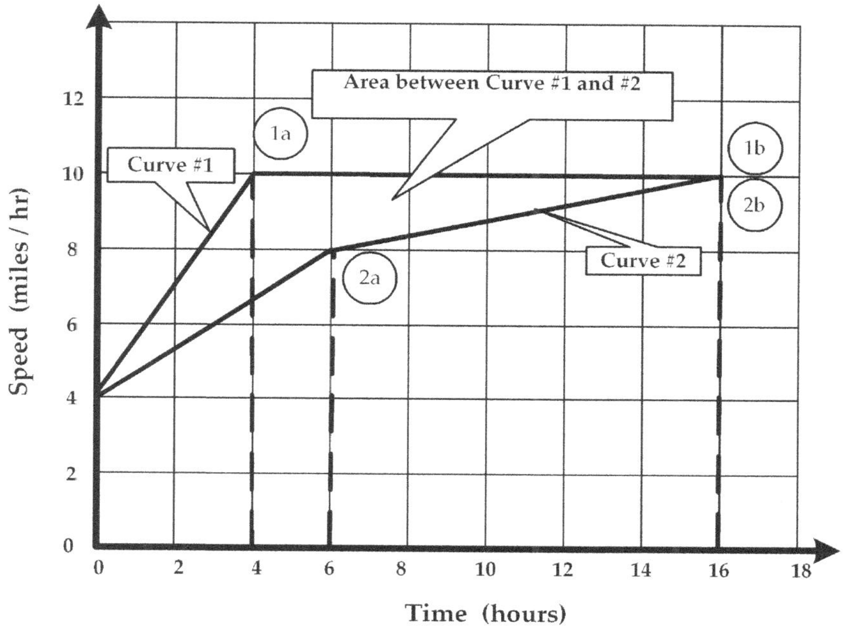

What is the area between Curve #1 and Curve #2 ?
The difference in distance travel between Car #1 and Car #2… (148 – 126 = 22 miles)

So how does this relate to entropy?

Let's go back to our origin question?
"How is the value 1.7577 (Btu / lbm-R) calculated ?"

Start by collecting the temperature and thermal energy data along two different paths of a medium like water. The data should look like the data from the previous automotive example (the distance and speed data from two cars). One set of data is collected along a constant pressure path while the other is collected along a constant volume path.

Water / Steam Data

Temp	Pressure	Volume	Entropy	Int. Energy	Enthalpy
°R	psia	ft³/lbm	Btu/lbm-°R	Btu/lbm	Btu/lbm
671.64	14.7	26.795	1.7577	1078.1	1151

Constant Pressure @ 14.7 psia

Temp	Pressure	T-Energy	???
°R	psia	Btu/lbm	???
491.69	14.7	0.00	???
500.00	14.7	8.41	???
600.00	14.7	108.42	???
671.64	14.7	180.30	???
671.64	14.7	1151.00	???

Constant Volume @ 26.795 ft3/lbm

Temp	Volume	T-Energy	???
°R	ft³/lbm	Btu/lbm	???
491.69	26.795	0.00	???
500	26.795	19.64	???
550	26.795	115.25	???
600	26.795	316.82	???
671.64	26.795	1078.10	???

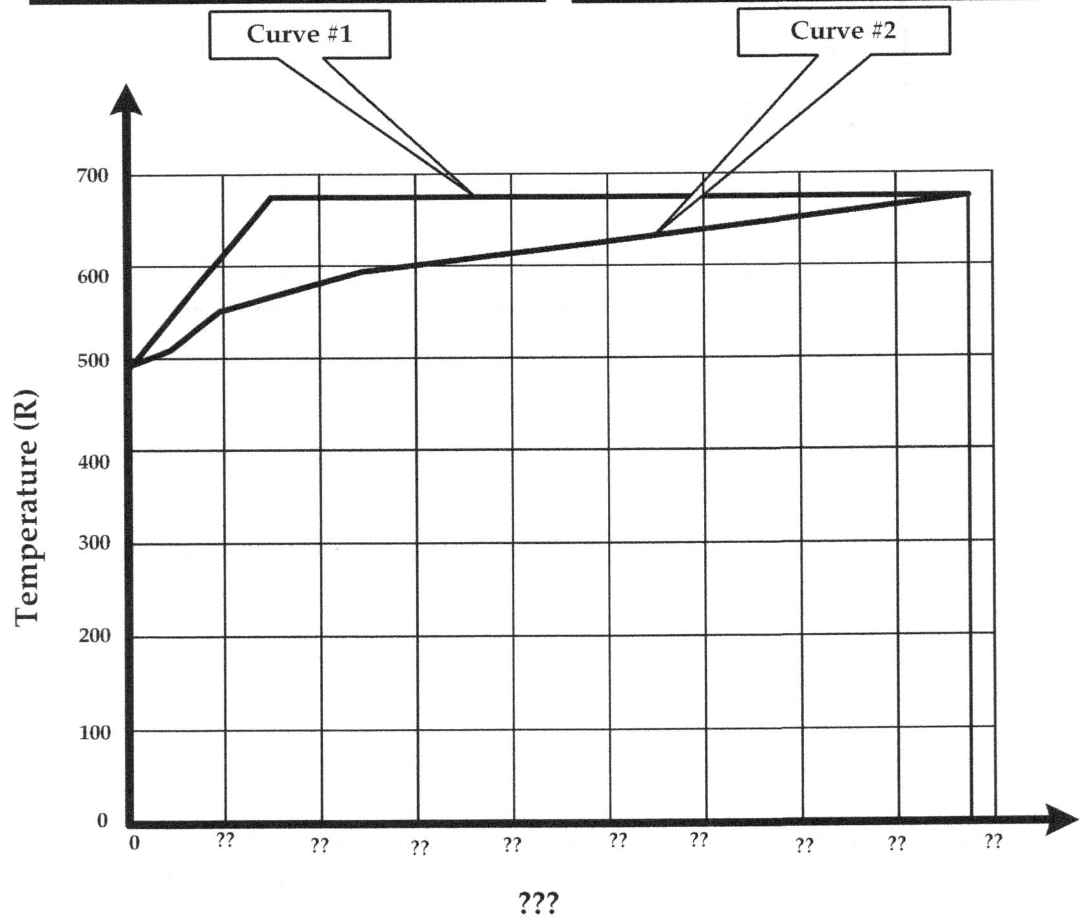

As "time" relates distance and speed together we need to ask.
What property relates temperature to thermal energy? ***Entropy***

Now substitute the automotive information from the previous car traveling example with the thermodynamics information as shown below

Automotive to Thermodynamics Conversion:

	Thermodynamics	Automotive
Curve #1	Constant pressure @ 14.7 psia	car #1
Curve #2	Constant volume @ 26.795 ft3/lbm	car #2
Area under curve	Thermal Energy	distance
y- axis	Temperature	speed
x-axis	Entropy	time

The automotive equations :

$$\Delta \text{ Time} = \frac{\Delta \text{ Distance}}{(\text{ Mean Speed })}$$

The thermodynamics equations :

$$\Delta \text{ Entropy} = \frac{\Delta \text{ Thermal Energy}}{(\text{ Mean Temperature })}$$

The automotive graph :

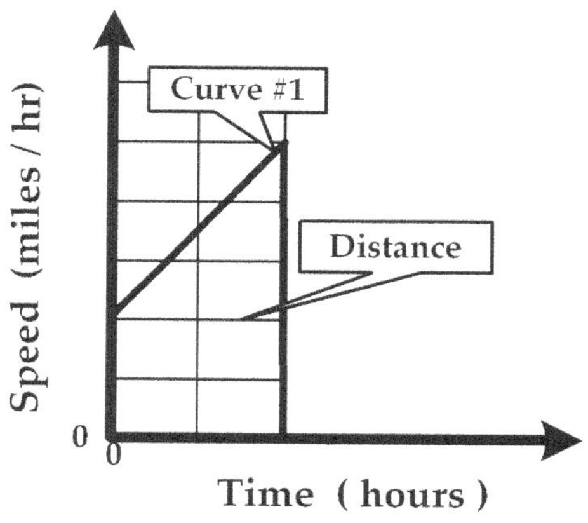

The thermodynamics graph :

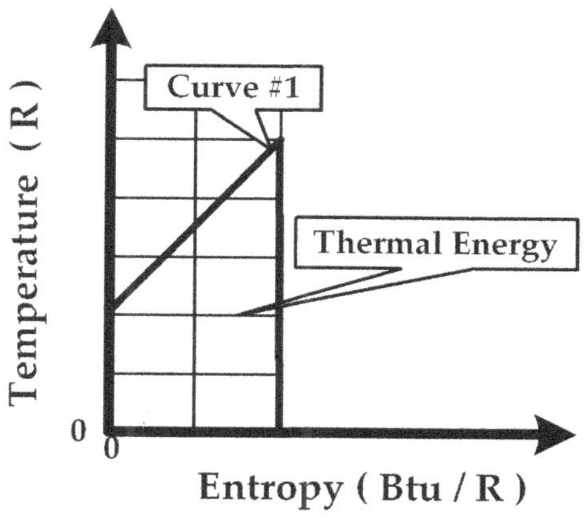

As the property *Time* is the property that relates distance to speed.

Entropy is a property of the medium that relates temperature to thermal energy.

The new set of equations for thermodynamics becomes:

<u>The thermodynamics equations :</u>

Assuming linear relationship betweem Temperature, Entropy and Thermal Energy

Tm = (Mean Temperature) = (1/2) (T 1 + T 2)

Δ Thermal Energy = (Mean Temperature) (Δ Entropy)

Δ TE (Btu) = (Tm)(°R)x (Δ S) (Btu/°R)

Δ Thermal Energy = (mass) (Δ specific thermal energy)

Δ TE (Btu) = (m)(lbm) x Δ te (Btu/lbm)

Δ specific thermal energy = (Mean Temperature) (Δ specific entropy)

Δ te (Btu/lbm) = (Tm)(°R)x (Δ s) (Btu/lbm-°R)

$$\Delta \text{ Entropy} = \frac{\Delta \text{ Thermal Energy}}{(\text{ Mean Temperature })} = \frac{(\text{ mass }) \times \Delta \text{ specific thermal energy}}{(\text{ Mean Temperature })}$$

$$\Delta \text{ S (Btu/°R)} = \frac{\Delta \text{ TE (Btu)}}{(\text{ Tm })(\text{°R})} = \frac{(\text{ m })(\text{lbm}) \times \Delta \text{te (Btu/lbm)}}{(\text{ Tm })(\text{°R})}$$

$$\Delta \text{ specific entropy} = \frac{\Delta \text{ specific thermal energy}}{(\text{ Mean Temperature })} \text{ (Btu / lbm -°R)}$$

$$\Delta \text{s} = \frac{\Delta \text{te (Btu/lbm)}}{(\text{ Tm })(\text{°R})} \text{ (Btu / lbm -°R)}$$

Before we use the new equations calculate the entropy value for each curve we need to add three more thermodynamics terms.

Enthalpy is the medium property that represents the thermal energy along a constant pressure path. (Curve #1…constant pressure @ 14.7 psia)

Internal Energy is the medium property that represents the thermal energy along a constant volume path. (Curve #2… constant volume @ 26.795 ft^3/lbm)

Pv work represents the thermal energy between the constant pressure path and the constant volume path. (***Pv work* = *Enthalpy - Internal Energy***)

Entropy and Enthalpy

Enthalpy is a medium property that represents the thermal energy along a constant pressure path.

Replacing the term "Thermal Energy" with "Enthalpy" a new set of equations can be established for a constant pressure path. Let's take our example above of Car 1 and change the variables around.

Area under the curve...distance with **Enthalpy**

Change the y-axis labeled speed with **Temperature**

Change the x-axis time with **Entropy**

The equation would be

Enthalpy (area under the curve) = Mean Temperature x Entropy
(along a constant *pressure* path)

The Entropy and Enthalpy equations :

Assuming linear relationship betweem Temperature, Entropy and Enthalpy

Tm = (Mean Temperature) = (1/2) (T 1 + T 2)

Δ Enthalpy = (Mean Temperature) (Δ Entropy)

$$\Delta H \text{ (Btu)} = (T_m)(^\circ R) \times (\Delta S) \text{ (Btu/}^\circ\text{R)}$$

Δ Enthalpy = (mass) (Δ specific thermal energy)

$$\Delta H \text{ (Btu)} = (m)(\text{lbm}) \times \Delta te \text{ (Btu/lbm)}$$

Δspecific enthalpy = (Mean Temperature) (Δspecific entropy)

$$\Delta h \text{ (Btu/lbm)} = (T_m)(^\circ R) \times (\Delta s)\text{(Btu/lbm-}^\circ\text{R)}$$

$$\Delta \text{ Entropy} = \frac{\Delta \text{ Enthalpy}}{(\text{Mean Temperature})} = \frac{(\text{mass}) \times \Delta \text{ specific enthalpy}}{(\text{Mean Temperature})}$$

$$\Delta S \text{ (Btu/}^\circ\text{R)} = \frac{\Delta H \text{ (Btu)}}{(T_m)(^\circ R)} = \frac{(m)(\text{lbm}) \times \Delta h \text{ (Btu/lbm)}}{(T_m)(^\circ R)}$$

$$\Delta \text{ specific entropy} = \frac{\Delta \text{ specific enthalpy}}{(\text{Mean Temperature})} \text{ (Btu / lbm - }^\circ\text{R)}$$

$$\Delta s = \frac{\Delta h \text{ (Btu/lbm)}}{(T_m)(^\circ R)} \text{ (Btu / lbm - }^\circ\text{R)}$$

Using the equations calculate the entropy for the constant pressure data.

Constant Pressure @ 14.7 psia

Temp	Pressure	Energy	Entropy
°R	psia	Btu/lbm	Btu/lbm-°R
491.69	14.7	0.00	s 0
500.00	14.7	8.41	s 1
600.00	14.7	108.42	s 2
671.64	14.7	180.30	s 3
671.64	14.7	1151.00	s 4

$$\Delta s = \frac{\Delta h}{Tm} = \frac{\Delta \text{ specific enthalpy}}{\text{Mean Temperature}} \quad \text{(Constant Pressure path)}$$

$$\Delta s0 = \frac{\Delta h}{Tm} = \frac{(0)}{(492)} = 0 \quad \text{(Btu/ lbm - R)}$$

$$\Delta s1 = \frac{\Delta h}{Tm} = \frac{(8.41 - 0)}{(0.5)(500+491.7)} = \frac{(8.41)}{(496)} = 0.017 \quad \text{(Btu/ lbm - R)}$$

$$\Delta s2 = \frac{\Delta h}{Tm} = \frac{(108.42 - 8.41)}{(0.5)(600 + 500)} = \frac{(100.01)}{(550)} = 0.182 \quad \text{(Btu/ lbm - R)}$$

$$\Delta s3 = \frac{\Delta h}{Tm} = \frac{(180.3 - 108.42)}{(0.5)(672 + 600)} = \frac{(71.88)}{(636)} = 0.113 \quad \text{(Btu/ lbm - R)}$$

$$\Delta s4 = \frac{\Delta h}{Tm} = \frac{(1151 - 108.42)}{(0.5)(672 + 672)} = \frac{(970.3)}{(672)} = 1.444 \quad \text{(Btu/ lbm - R)}$$

$s0 = 0$

$s1 = s0 + \Delta s1 = 0 + 0.017 = 0.017$ (Btu/ lbm - R)

$s2 = s1 + \Delta s2 = 0.017 + 0.182 = 0.199$ (Btu/ lbm - R)

$s3 = s2 + \Delta s3 = 0.199 + 0.113 = 0.312$ (Btu/ lbm - R)

$s4 = s3 + \Delta s4 = 0.312 + 1.444 = 1.756$ (Btu/ lbm - R)

Take the calculate entropy and update table. Also plot a graph showing the information

Water / Steam Data

Temp	Pressure	Volume	Entropy	Int. Energy	Enthalpy
°R	psia	ft³/lbm	Btu/lbm-°R	Btu/lbm	Btu/lbm
671.64	14.7	26.795	1.7577	1078.1	1151

Constant Pressure @ 14.7 psia

Temp	Pressure	Enthalpy	Calculated Entropy	Actual Entropy
°R	psia	Btu/lbm	Btu/lbm-°R	Btu/lbm-°R
491.69	14.7	0.00	0.000	0.0000
500.00	14.7	8.41	0.017	0.0169
600.00	14.7	108.42	0.199	0.1992
671.64	14.7	180.30	0.312	0.3124
671.64	14.7	1151.00	1.756	1.7577

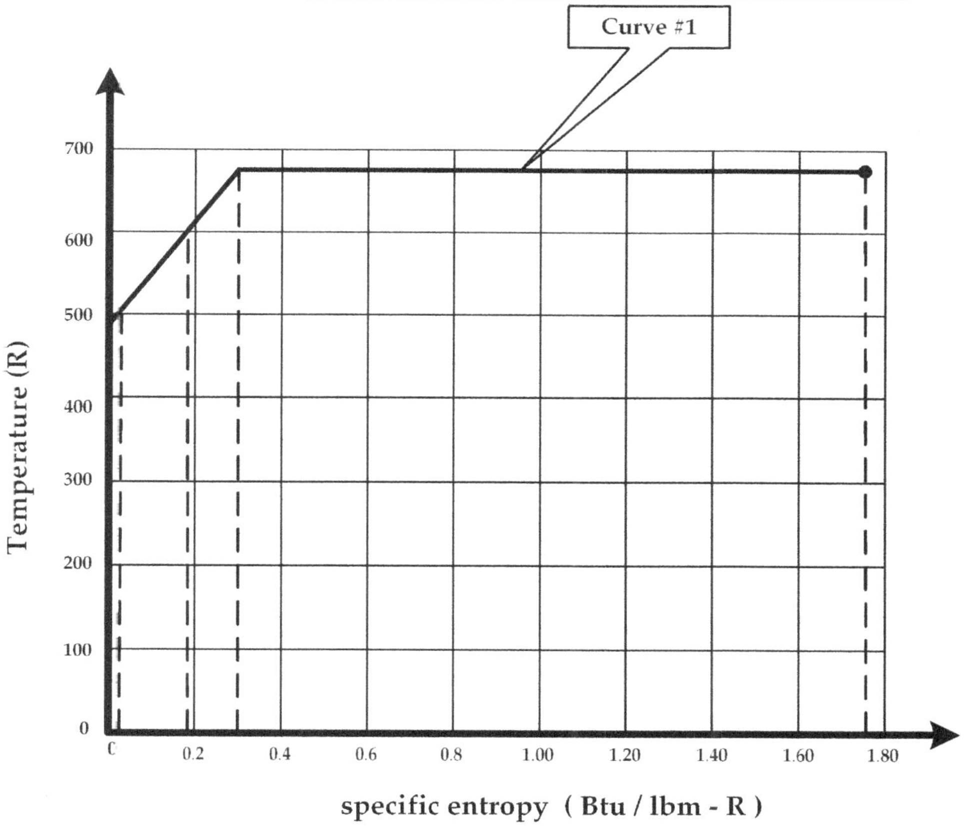

Reviewing the data from both tables we notice that the actual entropy data (Δs) and the calculated value in the constant pressure table are almost the same. The lines between each data points are linear (straight) for the calculated Δs values.

Entropy and Internal Energy

Internal Energy is a medium property that represents the thermal energy along a constant volume path.

Replacing the term "Thermal Energy" with "Internal Energy" a new set of equations can be established for a constant volume path.

Let's take our example above of Car 2 and change the variables around.

Area under the curve…distance with **Internal Energy**

Change the y-axis labeled speed with **Temperature**

Change the x-axis time with **Entropy**

The equation would be

Internal Energy (area under the curve) = Mean Temperature x Entropy
(along a constant *volume* path)

The Entropy and Internal Energy equations :

Assuming linear relationship betweem Temperature, Entropy and Internal Energy

Tm = (Mean Temperature) = (1/2) (T 1 + T 2)

Δ Internal Energy = (Mean Temperature) (Δ Entropy)

Δ U (Btu) = (Tm)(°R)x (Δ S) (Btu/°R)

Δ Thermal Energy = (mass) (Δ specific internal energy)

Δ U (Btu) = (m)(lbm) x Δ u (Btu/lbm)

Δ specific internal energy = (Mean Temperature) (Δ specific entropy)

Δ u (Btu/lbm) = (Tm)(°R)x (Δ s) (Btu/lbm-°R)

$$\Delta \text{ Entropy} = \frac{\Delta \text{ Internal Energy}}{(\text{Mean Temperature})} = \frac{(\text{mass}) \times \Delta \text{ specific internal energy}}{(\text{Mean Temperature})}$$

$$\Delta S \ (\text{Btu}/^\circ\text{R}) = \frac{\Delta U \ (\text{Btu})}{(\text{Tm})(^\circ\text{R})} = \frac{(\text{m})(\text{lbm}) \times \Delta u \ (\text{Btu/lbm})}{(\text{Tm})(^\circ\text{R})}$$

$$\Delta \text{ specific entropy} = \frac{\Delta \text{ specific internal energy}}{(\text{Mean Temperature})} \ (\text{Btu / lbm -}^\circ\text{R})$$

$$\Delta s = \frac{\Delta u \ (\text{Btu/lbm})}{(\text{Tm})(^\circ\text{R})} \ (\text{Btu / lbm -}^\circ\text{R})$$

Using the equations calculate the entropy for the constant volume data.

Constant Volume @ 26.795 ft^3/lbm

Temp	Volume	Energy	Entropy
°R	ft³/lbm	Btu/lbm	Btu/lbm-°R
491.69	26.795	0.00	s 0
500	26.795	19.64	s 1
550	26.795	115.25	s 2
600	26.795	316.82	s 3
671.64	26.795	1078.10	s 4

$$\Delta s = \frac{\Delta u}{Tm} = \frac{\Delta \text{ specific internal energy}}{\text{Mean Temperature}} \quad \text{(Constant Volume path)}$$

$$\Delta s0 = \frac{\Delta u}{Tm} = \frac{(0)}{(492)} = 0 \quad \text{(Btu/ lbm - R)}$$

$$\Delta s1 = \frac{\Delta u}{Tm} = \frac{(19.64 - 0)}{(0.5)(500+491.7)} = \frac{(19.64)}{(496)} = 0.040 \quad \text{(Btu/ lbm - R)}$$

$$\Delta s2 = \frac{\Delta u}{Tm} = \frac{(115.25 - 19.64)}{(0.5)(550 + 500)} = \frac{(95.61)}{(525)} = 0.182 \quad \text{(Btu/ lbm - R)}$$

$$\Delta s3 = \frac{\Delta u}{Tm} = \frac{(316.82 - 115.25)}{(0.5)(600 + 550)} = \frac{(201.57)}{(575)} = 0.351 \quad \text{(Btu/ lbm - R)}$$

$$\Delta s4 = \frac{\Delta u}{Tm} = \frac{(1151 - 108.42)}{(0.5)(672 + 600)} = \frac{(761.28)}{(636)} = 1.197 \quad \text{(Btu/ lbm - R)}$$

$s0 = 0$

$s1 = s0 + \Delta s1 = 0 + 0.040 = 0.040$ (Btu/ lbm - R)

$s2 = s1 + \Delta s2 = 0.040 + 0.182 = 0.222$ (Btu/ lbm - R)

$s3 = s2 + \Delta s3 = 0.222 + 0.351 = 0.573$ (Btu/ lbm - R)

$s4 = s3 + \Delta s4 = 0.573 + 1.197 = 1.770$ (Btu/ lbm - R)

Take the calculate entropy and update table. Also plot a graph showing the information

Water / Steam Data

Temp	Pressure	Volume	Entropy	Int. Energy	Enthalpy
°R	psia	ft³/lbm	Btu/lbm-°R	Btu/lbm	Btu/lbm
671.64	14.7	26.795	1.7577	1078.1	1151

Constant Volume @ 26.795 ft³ / lbm

Temp	Volume	Int Energy	Calculated Entropy	Actual Entropy
°R	ft³/lbm	Btu/lbm	Btu/lbm-°R	Btu/lbm-°R
491.69	26.795	0.00	0.000	0.0000
500.00	26.795	19.64	0.040	0.0406
550.00	26.795	115.25	0.222	0.2220
600.00	26.795	316.82	0.573	0.5708
671.64	26.795	1078.10	1.770	1.7577

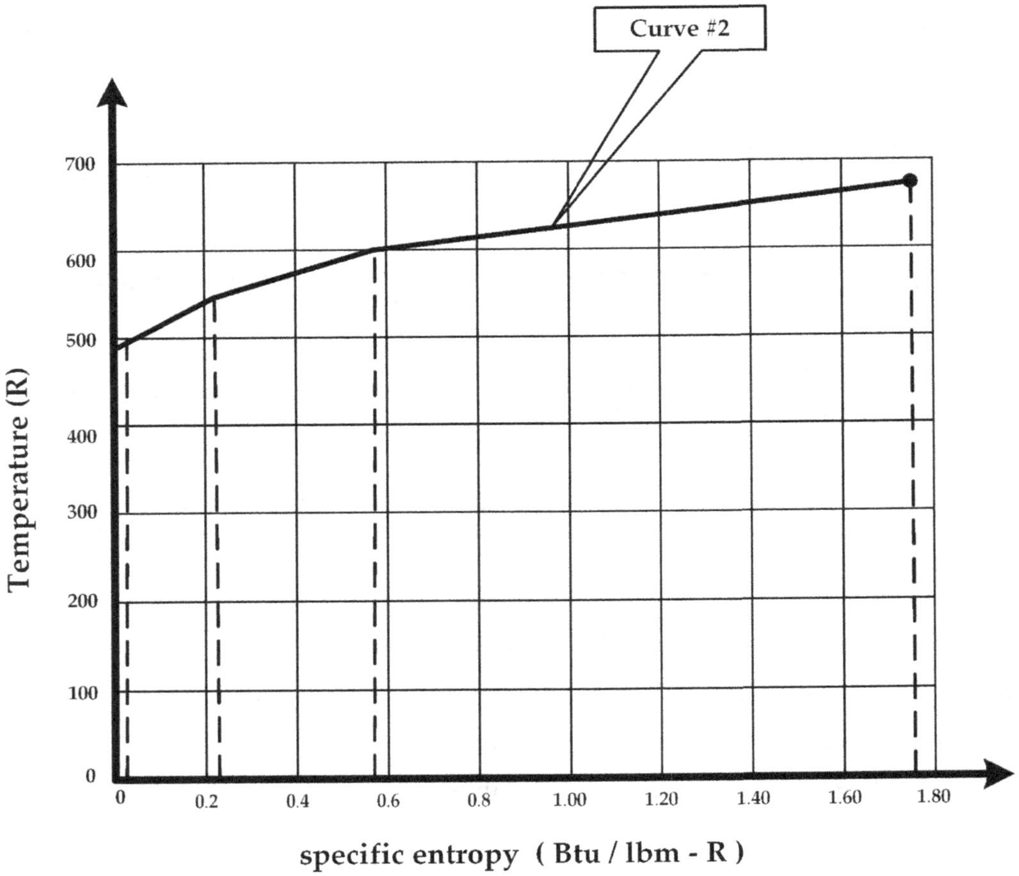

At this time take a look at the entropy data in the table above constant volume of 26.795 ft^3/lbm. Looking at the data you may ask, "why is the calculated entropy different than the actual entropy in the constant specific volume table?". I'm glad you asked. Remember earlier we *assumed* the line between two data points to be linear (straight) to calculate Δs. Reviewing the data from constant pressure table we notice that the actual entropy data (Δs) and the calculated value were almost identical. While the data Δs in the constant volume table doesn't match.

Conclusion, the data in the constant volume table needs to be broken down into smaller lines to find an accurate ΔS change in entropy. Therefore to increase the accuracy of the data we need to calculate the entropy for each degree of temperature. Instead of going from 600 to 672 we should have collected data from 600 to 601, 601 to 602 and etc. Below is data and calculation of entropy between 600 R and 605 R at a constant specific volume of 26.795 ft^3/lbm.

Constant Volume @ 26.795 ft^3 / lbm

Temp °R	Volume ft³/lbm	Int Energy Btu/lbm	Entropy Btu/lbm-°R
600.00	26.795	316.82	s 0
601.00	26.795	322.74	s 1
602.00	26.795	328.76	s 2
603.00	26.795	334.87	s 3
604.00	26.795	341.08	s 4
605.00	26.795	347.40	s 5

$$\Delta s0 = \frac{\Delta u}{Tm} = 0.57075$$

$$\Delta s1 = \frac{\Delta u}{Tm} = \frac{(322.75 - 316.82)}{(O.5)(601+600)} = \frac{(5.92)}{(600.5)} = 0.00986 \quad \textbf{(Btu/ lbm - R)}$$

$$\Delta s2 = \frac{\Delta u}{Tm} = \frac{(328.76 - 322.74)}{(O.5)(602+601)} = \frac{(6.02)}{(601.5)} = 0.01000 \quad \textbf{(Btu/ lbm - R)}$$

$$\Delta s3 = \frac{\Delta u}{Tm} = \frac{(334.87 - 328.76)}{(O.5)(603+602)} = \frac{(6.11)}{(602.5)} = 0.01014 \quad \textbf{(Btu/ lbm - R)}$$

$$\Delta s4 = \frac{\Delta u}{Tm} = \frac{(341.08 - 334.87)}{(O.5)(604+603)} = \frac{(6.21)}{(603.5)} = 0.01029 \quad \textbf{(Btu/ lbm - R)}$$

$$\Delta s5 = \frac{\Delta u}{Tm} = \frac{(347.4 - 341.08)}{(O.5)(605+604)} = \frac{(6.32)}{(604.5)} = 0.01045 \quad \textbf{(Btu/ lbm - R)}$$

Constant Volume @ 26.795 ft^3 / lbm

s0 = 0.57075 (Btu/ lbm - R)

s1 = s0 + Δ s1 = 0.57075 + 0.00986 = 0.58061 (Btu/ lbm - R)

s2 = s1 + Δ s2 = 0.58061 + 0..01000 = 0.59061 (Btu/ lbm - R)

s3 = s2 + Δ s3 = 0.59061 + 0.1014 = 0.60075 (Btu/ lbm - R)

s4 = s3 + Δ s4 = 0.60075 + 0.01029 = 0.61104 (Btu/ lbm - R)

s4 = s3 + Δ s4 = 0.61104 + 0.01045 = 0.62149 (Btu/ lbm - R)

			Calculated	Actual
Temp	Volume	Int Energy	Entropy	Entropy
°R	ft³/lbm	Btu/lbm	Btu/lbm-°R	Btu/lbm-°R
600.00	26.795	316.82	0.57075	0.57075
601.00	26.795	322.74	0.58061	0.58061
602.00	26.795	328.76	0.59061	0.59061
603.00	26.795	334.87	0.60075	0.60075
604.00	26.795	341.08	0.61104	0.61105
605.00	26.795	347.40	0.62149	0.62149

Collect all the ΔU and ΔT value between the temperatures 600 R and 672 R in one degree intervals to obtain an accurate value of ΔS.

Pv Work = Enthalpy - Internal Energy

Enthalpy is the medium property that represents the thermal energy along a constant pressure path. (Curve #1…constant pressure @ 14.7 psia)

Internal Energy is the medium property that represents the thermal energy along a constant volume path. (Curve #2… constant volume @ 26.795 ft³/lbm)

What is the area between the two lines?

Pv work represents the thermal energy between the constant pressure path and the constant volume path. (***Pv work* = *Enthalpy - Internal Energy***)

From the definition in most thermodynamics book for enthalpy

Enthalpy = Internal Energy + PV

$$H = U + PV = (m)(h) \quad \text{(Btu)}$$

For specific enthalpy the equation is:

$$h = (u + Pv)\ \text{(Btu/lbm)}$$

Therefore the area between the Enthalpy cuve and the Internal Energy

PV = Enthalpy - Internal Energy

$$PV = H - U = (m)(Pv) \quad \text{(Btu)}$$

$$Pv = h - u \quad \text{(Btu/lbm)}$$

What is wrong with these equation ? the units, they don't match

$$Pv\ (\text{psia})\ (\text{ft3 / lbm}) \neq h\ (\text{Btu/lbm}) - u\ (\text{Btu/lbm})$$

Take a look at the intersection of two curves described earlier:

Water / Steam Data

Temp	Pressure	Volume	Entropy	Int. Energy	Enthalpy
°R	psia	ft³/lbm	Btu/lbm-°R	Btu/lbm	Btu/lbm
671.64	14.7	26.795	1.7577	1078.1	1151

$$(14.7\ \text{psia})\ (26.795\ \text{ft3 / lbm}) \neq (1151\ \text{Btu/lbm}) - (1078.1\ \text{Btu/lbm})$$

Convert (psia) (ft3 / lbm) to (Btu / lbm)

$$1\ (\text{Btu/lbm}) = \frac{1\ (\text{lbf})}{\text{in}^2} \times \frac{1\ (\text{ft}^3)}{(\text{lbm})} \times \frac{1\ (\text{Btu})}{778\ (\text{ft -lbf})} \times \frac{144\ (\text{in}^2)}{1\ (\text{ft}^2)}$$

$$1\ (\text{Btu/lbm}) = 0.185\ (\text{psia})\ (\text{ft}^3 / \text{lbm})$$

$$H = U + (0.185)\ P\ V = (m)(h)$$

For specific enthalpy the equation is:

$$h = u + (0.185)\ P\ v$$

$$(0.185)\ P\ v = (h) - (u)$$

$$(0.185)(14.7\ \text{psia})\ (26.795\ \text{ft3 / lbm}) = (1151\ \text{Btu/lbm}) - (1078.1\ \text{Btu/lbm})$$

$$(72.9\ \text{Btu/lbm}) = (72.9\ \text{Btu/lbm})$$

Last Reminder: UNITS MUST MATCH

Looking at the complete picture...
To illustrate this point lets look at a typical water / steam data points. Update the data collected at a constant pressure of 14.7 psia and at a constant volume of 26.795 ft³/lbm. Plot the data points from the two curves on the same T-s diagram.

Water / Steam Data	Constant Pressure @ 14.7 psia					
Temp °R	Pressure psia	Volume ft³/lbm	Entropy Btu/lbm-°R	Int. Energy Btu/lbm	Enthalpy Btu/lbm	Pv work Btu/lbm
491.69	14.7	0	0.0000	0.00	0.00	0.00
500	14.7	0.01602	0.0169	8.37	8.41	0.04
600	14.7	0.01629	0.1992	108.38	108.42	0.04
671.64	14.7	0.01671	0.3124	180.25	180.30	0.05
671.64	14.7	26.795	1.7577	1078.10	1151.00	72.87

Water / Steam Data	Constant Volume @ 26.795 ft³ / lbm					
Temp °R	Pressure psia	Volume ft³/lbm	Entropy Btu/lbm-°R	Int. Energy Btu/lbm	Enthalpy Btu/lbm	Pv work Btu/lbm
491.69	0	26.795	0.0000	0.00	0.00	0.00
500	0.12331	26.795	0.0406	19.64	20.25	0.61
550	0.70631	26.795	0.2220	115.25	118.76	3.50
600	2.9176	26.795	0.5708	316.82	331.30	14.46
671.64	14.7	26.795	1.7577	1078.10	1151.00	72.87

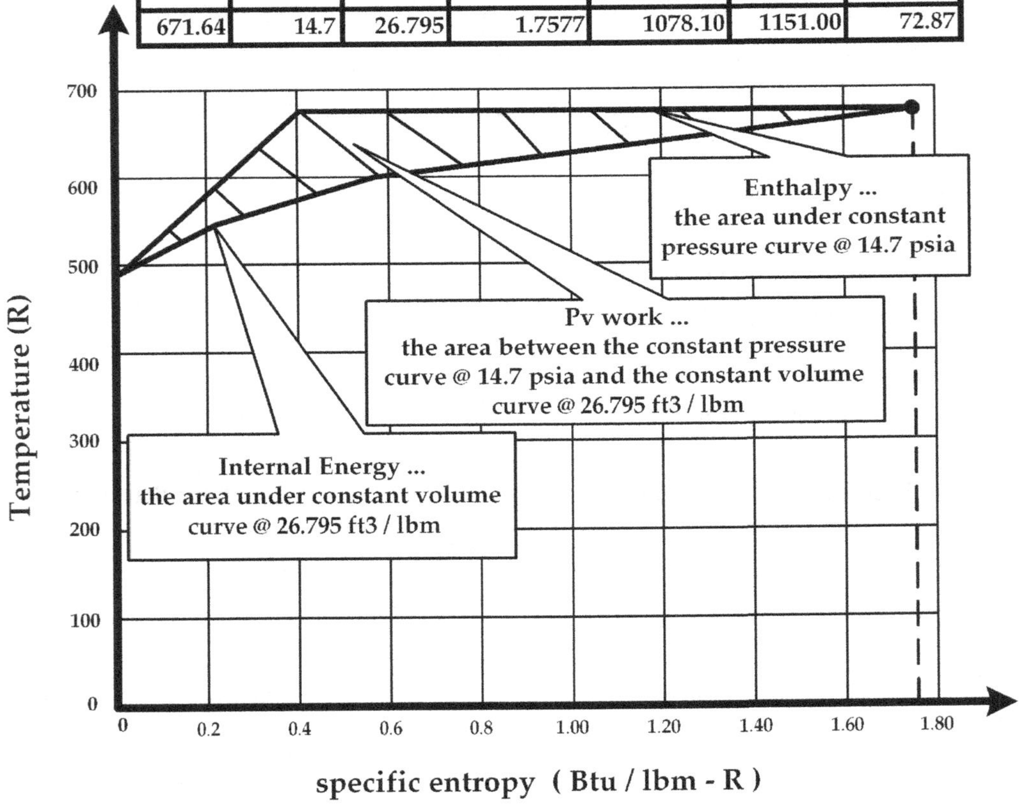

Observation from the data and graph:

1. Both lines started at 492 R and S=0
2. Both ended at 672 R and S= 1.7577.
3. One curve follows a constant pressure of 14.7 psia
4. The other curve follows constant volume of 26.795 ft^3/lbm

Updating the two tables:

Note one other item about entropy. The intersection of the two curves intersect at a Temperature of 671.64 °R and entropy of 1.7577 Btu/lbm-°R. Therefore, by knowing the temperature and entropy along the constant pressure and constant volume curves that intersect the point we are to define the properties of the medium.

FYI… you just did an integral in calculus, by solving the area under a curve.(trapezoid rule)

2.2 Enthalpy work and enthalpy heat

The terms and "enthalpy work" and "enthalpy heat" are new terms in the study of thermodynamics and represent the available and unavailable enthalpy to do work of a medium along a constant pressure.

Enthalpy = enthalpy work + enthalpy heat

Enthalpy = enthalpy available to do work + enthalpy unavailable to do work

The terms enthalpy work and enthalpy heat came from the study of "exergy" and the famous equation "G = H –T S" called the Gibbs function. This function on free energy was developed by Willard Gibbs in the 1870's. Please take the time to read about Willard Gibbs, an American mathematical physicist and a legend in the field of thermodynamics. The only problem I had with the explanation of the Gibbs function was the term "free energy". Anyone studying thermodynamics soon realizes there is no free energy.

The best way to explain the terms "enthalpy work" and "enthalpy heat" is with an example. From our definition of enthalpy work and enthalpy heat let's use the same data and graph from our previous water and steam example.

.

Water / Steam Data — Constant Pressure @ 14.7 psia

Temp	Pressure	Volume	Entropy	Int. Energy	Enthalpy	Pv work
°R	psia	ft^3/lbm	Btu/lbm-°R	Btu/lbm	Btu/lbm	Btu/lbm
491.69	14.7	0	0.0000	0.00	0.00	0.00
500	14.7	0.01602	0.0169	8.37	8.41	0.04
600	14.7	0.01629	0.1992	108.38	108.42	0.04
671.64	14.7	0.01671	0.3124	180.25	180.30	0.05
671.64	14.7	26.795	1.7577	1078.10	1151.00	72.87

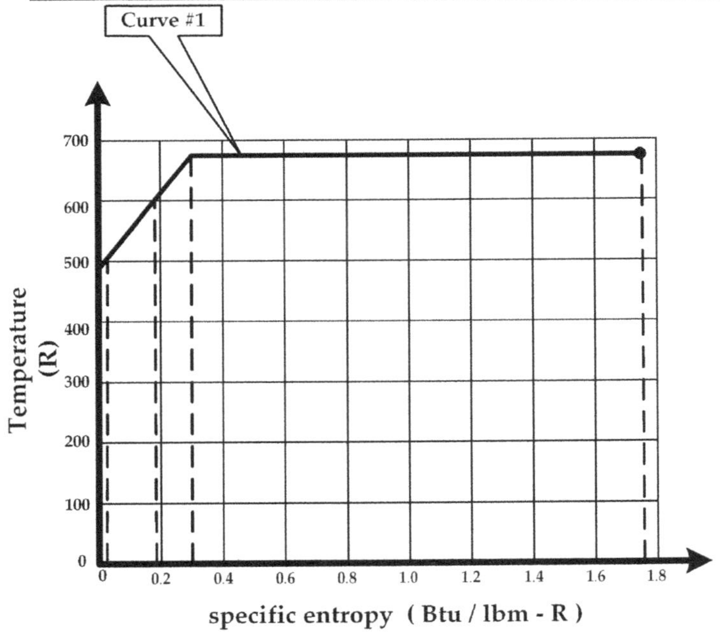

Take the constant pressure data and apply it to the Gibbs function "G = H –T S" is defined as:

Free energy = Enthalpy – (Temperature x Entropy)

By rearranging the equation to define the enthalpy "H = G + T S" :

Enthalpy = Free energy + (Temperature x Entropy)

By redefining the variables G = Hw and TS = Hq.
The equation "H = G + T S" becomes "H = Hw + Hq" to obtain the terms "enthalpy work" and "enthalpy heat".

Enthalpy = Enthalpy Work + Enthalpy Heat
Enthalpy = (mass) x (specific enthalpy)
Specific enthalpy = specific enthalpy work + specific enthalpy heat

From the last section we defined "Enthalpy" as the area under a constant pressure curve on a T-s diagram.

The term "Enthalpy work" is defined as the part of enthalpy that is available to do work. The ***area of the curve above*** the reference temperature (491.69 °R). The symbol for enthalpy work is "Hw" and units of "Btu". And the symbol for specific enthalpy work is "hw" and units of "Btu / lbm".

"Enthalpy heat" is the part of enthalpy unavailable to do work The ***area of the curve below*** the reference temperature (491.69 °R). The symbol for enthalpy heat is "Hq" and with the units of "Btu". The symbol for specific enthalpy heat is "hq" and with the units of "Btu/lbm".

The "enthalpy work" and "enthalpy heat" are the available and unavailable enthalpy of a medium.

Enthalpy = enthalpy available to do work + enthalpy unavailable to do work

Enthalpy = enthalpy work + enthalpy heat

Enthalpy....H = Hw + Hq = (m) (h) (Btu)

Specific enthalpy... h = hw + hq (Btu/lbm)

Note the reference temperature 491.69 °R is different for each medium. At this temperature the entropy, internal energy and enthalpy are zero. If look though any thermodynamic book you may see a reference temperature for ammonia at 351.89 (°R) or 420 (°R). The refrigerant R 410 may be at 360 (°R) or 420 (°R) depending on book.

Using the previous constant pressure graph, enthalpy work and enthalpy heat can be displayed on the T-s diagram below:

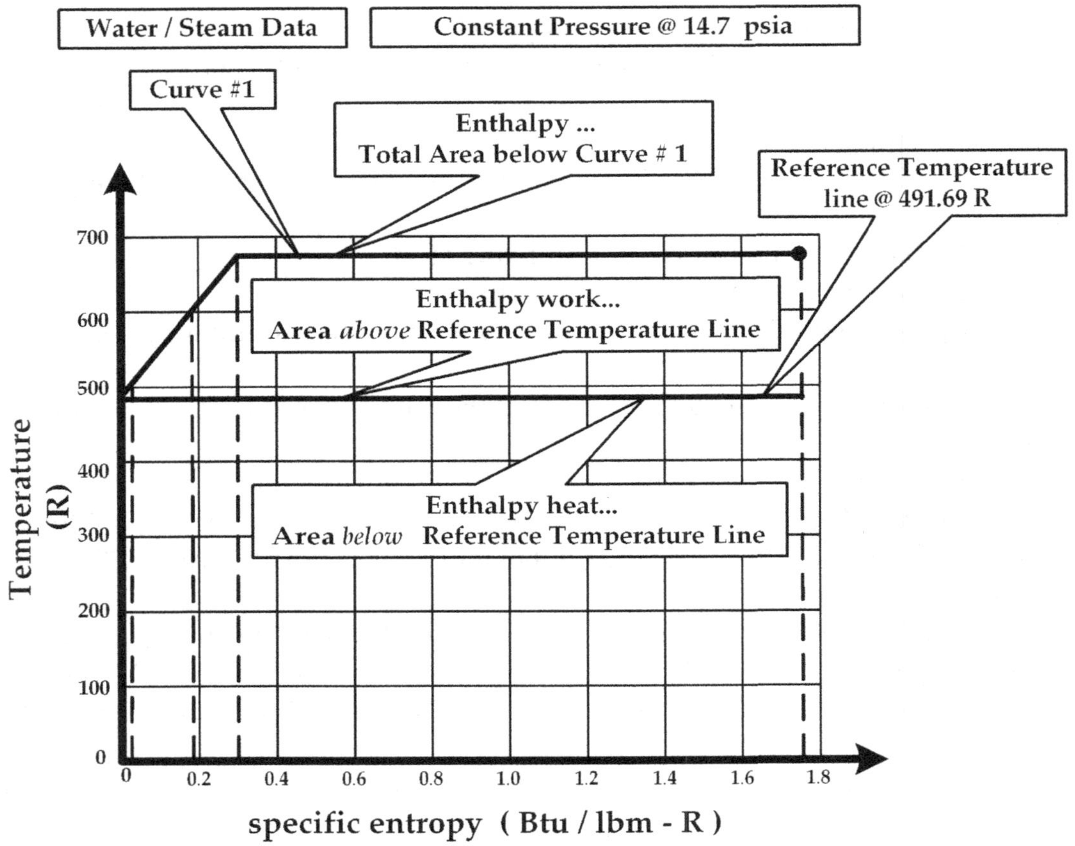

From the graph:

1. The enthalpy is equal to the total area under the constant pressure curve # 1.

Next divide the enthalpy (area under curve) into two sections:

a. The "enthalpy heat" is the area of under the curve below the temperature 491.69 °R.
Enthalpy Heat is simply Entropy times Reference Temperature

b. The "enthalpy work" is the remaining area under the curve or simply:
enthalpy work is equal to enthalpy (total area) – enthalpy heat (area below 491.69 °R)

Looking at the graph above the calculation for the enthalpy heat is simply entropy times Reference Temperature. Again the reference temperature is 491.68 °R. From Table below calculate hq and hw using the following equation:

1. Enthalpy…"h" supplied by table
2. Enthalpy heat "hq = Tr x s" = Reference Temperate times entropy
3. Enthalpy work "hw = h – hq" = enthalpy minus enthalpy heat

Constant Pressure @ 14.7 psia **Reference Temperature = 491.69 R**

Temp	Pressure	Volume	Entropy	Int. Energy	Enthalpy	hq	hw	Pv work
°R	psia	ft³/lbm	Btu/lbm-°R	Btu/lbm	Btu/lbm	Btu/lbm	Btu/lbm	Btu/lbm
491.69	14.7	0	0.0000	0.00	0.00	hq0	hw 0	0.00
500	14.7	0.016019	0.0169	8.37	8.41	hq1	hw1	0.04
600	14.7	0.016294	0.1992	108.38	108.42	hq2	hw2	0.04
671.64	14.7	0.016714	0.3124	180.25	180.30	hq3	hw3	0.05
671.64	14.7	26.795	1.7577	1078.10	1151.00	hq4	hw4	72.87

Calculating: **hq = specific enthalpy heat** **hw = specific enthalpy work**

Δ hq = (Tr) x Δ s = (Reference Temperature) x Δ specific entropy

Δ hw = Δ h - Δ hq = Δ specific enthalpy - Δ specific enthalpy heat

Using a starting reference temperature of 491.69 R and a "0" entropy the equations become:

hq = (Tr) x (s)

hw = (h) - (hq)

hq0 = (491.69) x (0) = 0 (Btu / lbm)

hw0 = (0) - (0) = 0 (Btu / lbm)

hq1 = (491.69) x (0.0169) = 8.30 (Btu / lbm)

hw1 = (8.41) - (8.30) = 0.11 (Btu / lbm)

hq2 = (491.69) x (0.1992) = 97.95 (Btu / lbm)

hw2 = (108.42) - (97.95) = 10.47 (Btu / lbm)

hq3 = (491.69) x (0.0169) = 153.59 (Btu / lbm)

hw3 = (180.25) - (153.59) = 26.71 (Btu / lbm)

hq4 = (491.69) x (1.7577) = 864.24 (Btu / lbm)

hw4 = (1151.00) - (864.24) = 286.76 (Btu / lbm)

Results from calculations:

Temp	Pressure	Volume	Entropy	Int. Energy	Enthalpy	hq	hw	Pv work
°R	psia	ft³/lbm	Btu/lbm-°R	Btu/lbm	Btu/lbm	Btu/lbm	Btu/lbm	Btu/lbm
491.69	14.7	0	0.0000	0.00	0.00	0.00	0.00	0.00
500	14.7	0.016019	0.0169	8.37	8.41	8.30	0.11	0.04
600	14.7	0.016294	0.1992	108.38	108.42	97.95	10.47	0.04
671.64	14.7	0.016714	0.3124	180.25	180.30	153.59	26.71	0.05
671.64	14.7	26.795	1.7577	1078.10	1151.00	864.24	286.76	72.87

Observation from this data:

1. Verify the data Point 4… hw = 286.5 (Btu/lbm) using:

hw = Mean Temperature times the entropy change minus enthalpy heat
hw= h- hq = (Tm x s) - hq = (0.5)(T1+ T2) (s) - hq
hw = (0.5)(671.6 +492) (0.3124) + (671.6)(1.4453) - 864.5 = 288.0 (Btu/lbm

hw = Area of triangle and rectangle
hw= 0.5 (T2 –T1) (s) + (T2 – T1) x s
hw = (0.5) (671.6 -492) (0.3124) + ((671.6-492)(1.4453) = 287.7 (Btu/lbm)

2. The reference temperature 491.69 °R used for the calculation of enthalpy work and enthalpy heat is the temperature where the enthalpy and entropy data starts or when entropy and enthalpy are "0" … zero.

3. Latent heat (671.64 °R water to 671.64 °R steam)

Take a glance at the latent energy change in enthalpy, enthalpy work and enthalpy heat:

The change in enthalpy is… 1151 – 180.3 = 970.7 (Btu/lbm)

The change in enthalpy heat is… 864.24 – 153.59 = 710.65 (Btu/lbm)

The change in enthalpy work is… 286.76 – 26.71 = 260.05 (Btu/lbm)

Notice for an increase in enthalpy of 970.7 (Btu/lbm) we only gained 260.05 (Btu/lbm) of energy to do work.

Energy is not free

Here is data collected at a constant pressure and assume the entropy data was not in hand, how would hq and hw be calculated ?

Constant Pressure @ 14.7 psia **Reference Temperature = 491.69 R**

Temp	Pressure	Volume	Entropy	Int. Energy	Enthalpy	hq	hw	Pv work
°R	psia	ft³/lbm	Btu/lbm-°R	Btu/lbm	Btu/lbm	Btu/lbm	Btu/lbm	Btu/lbm
491.69	14.7	0	[redacted]	0.00	0.00	hq0	hw 0	0.00
500	14.7	0.016019	[redacted]	8.37	8.41	hq1	hw1	0.04
600	14.7	0.016294	[redacted]	108.38	108.42	hq2	hw2	0.04
671.64	14.7	0.016714	[redacted]	180.25	180.30	hq3	hw3	0.05
671.64	14.7	26.795	[redacted]	1078.10	1151.00	hq4	hw4	72.87

Calculating: hq = specific enthalpy heat
hw = specific enthalpy work " Without entropy data "

$\Delta hq = (Tr) \times \Delta s$ = (Reference Temperature) x Δspecific entropy

$\Delta hw = \Delta h - \Delta hq$ = Δ specific enthalpy - Δspecific enthalpy heat

Replacing: Δs for $\dfrac{\Delta h}{(Tm)} = \dfrac{\Delta \text{ specific enthalpy}}{(\text{Mean Temperature})}$

$$\Delta hq = (Tr) \times \frac{\Delta h}{(Tm)} \qquad \Delta hw = \Delta h - \Delta hq$$

<u>Assuming data is linear between points :</u>

hq0 = 0 (Btu / lbm) hw0 = 0 (Btu / lbm)

$$hq1 - hq0 = (491.69) \times \frac{(8.41 - 0)}{(0.5)(500 + 491.69)} = 8.30 \text{ (Btu / lbm)}$$

$$hw1 - hw0 = (8.41 - 0) - (8.30) = 0.11 \text{ (Btu / lbm)}$$

$$hq2 - hq1 = (491.69) \times \frac{(108.42 - 8.41)}{(0.5)(600 + 500)} = 89.41 \text{ (Btu / lbm)}$$

$$hw2 - hw1 = (108.42 - 8.41) - (89.41) = 10.60 \text{ (Btu / lbm)}$$

$$hq3 - hq2 = (491.69) \times \frac{(180.25 - 108.42)}{(0.5)(671.64 + 600)} = 55.55 \text{ (Btu / lbm)}$$

$$hw3 - hw2 = (180.25 - 108.42) - (55.55) = 16.28 \text{ (Btu / lbm)}$$

$$hq4 - hq3 = (491.69) \times \frac{(1151.0 - 180.25)}{(0.5)(671.64 + 671.64)} = 710.61 \text{ (Btu / lbm)}$$

$$hw4 - hw3 = (1151.0 - 180.25) - (710.61) = 260.14 \text{ (Btu / lbm)}$$

Calculating:

hq = specific enthalpy heat

hw = specific enthalpy work

hq (n) = hq (n-1) + Δ hq

hw (n) = hw (n-1) + Δ hw

hq0 = 0 (Btu / lbm)

hw0 = 0 (Btu / lbm)

hq1 = hq0 + (hq1- hq0) = (0) + (8.30) = 8.30 (Btu / lbm)

hw1 = hw0 + (hw1- hw0) = (0) + (0.11) = 0.11 (Btu / lbm)

hq2 = hq1 + (hq2- hq1) = (8.30) + (89.41) = 97.71 (Btu / lbm)

hw2 = hw1 + (hw2- hw1) = (0.11) + (10.60) = 10.71 (Btu / lbm)

hq3 = hq2 + (hq3- hq2) = (97.71) + (55.55) = 153.26 (Btu / lbm)

hw3 = hw2 + (hw3- hw2) = (10.71) + (16.28) = 26.99 (Btu / lbm)

hq4 = hq3 + (hq4- hq3) = (153.26) + (710.61) = 863.87 (Btu / lbm)

hw4 = hw3 + (hw4- hw3) = (26.99) + (260.14) = 286.13 (Btu / lbm)

Compare results from previous calculations:

Temp	Pressure	Volume	Entropy	Int. Energy	Enthalpy	hq	hw	Pv work
°R	psia	ft³/lbm	Btu/lbm-°R	Btu/lbm	Btu/lbm	Btu/lbm	Btu/lbm	Btu/lbm
491.69	14.7	0	0.0000	0.00	0.00	0.00	0.00	0.00
500	14.7	0.016019	0.0169	8.37	8.41	8.30	0.11	0.04
600	14.7	0.016294	0.1992	108.38	108.42	97.95	10.47	0.04
671.64	14.7	0.016714	0.3124	180.25	180.30	153.59	26.71	0.05
671.64	14.7	26.795	1.7577	1078.10	1151.00	864.24	286.76	72.87

2.3 Internal energy heat and internal energy work

The terms "internal energy work" and "internal energy heat" are also new terms in the study of thermodynamics and represent the available and unavailable internal energy to do work of a medium along a constant volume.

Internal energy = internal energy work + internal energy heat

Internal energy = internal energy available to do work + internal energy unavailable to do work

The terms internal energy work and internal energy heat came from the famous equation "F = U –T S" called the Helmholtz function. This theory on free energy was developed by Hermann von Helmholtz in the 1870's. Please take the time to read about Hermann von Helmholtz, an American mathematical physicist and a legend in the field of thermodynamics.

The best way to explain the terms "internal energy work" and "internal energy heat" is with an example. Remember the previous data from a typical water steam table at a constant volume of 26.795 ft³/lbm. Notice the data below starts from a reference temperature of 491.69 °R which is typical for water and steam tables. At this temperature the entropy, internal energy and enthalpy are zero.

Water / Steam Data | Constant Volume @ 26.795 ft³ / lbm

Temp	Pressure	Volume	Entropy	Int. Energy	Enthalpy	Pv work
°R	psia	ft³/lbm	Btu/lbm-°R	Btu/lbm	Btu/lbm	Btu/lbm
491.69	0	26.795	0.0000	0.00	0.00	0.00
500	0.12331	26.795	0.0406	19.64	20.25	0.61
550	0.70631	26.795	0.2220	115.25	118.76	3.50
600	2.9176	26.795	0.5708	316.82	331.30	14.46
671.64	14.7	26.795	1.7577	1078.10	1151.00	72.87

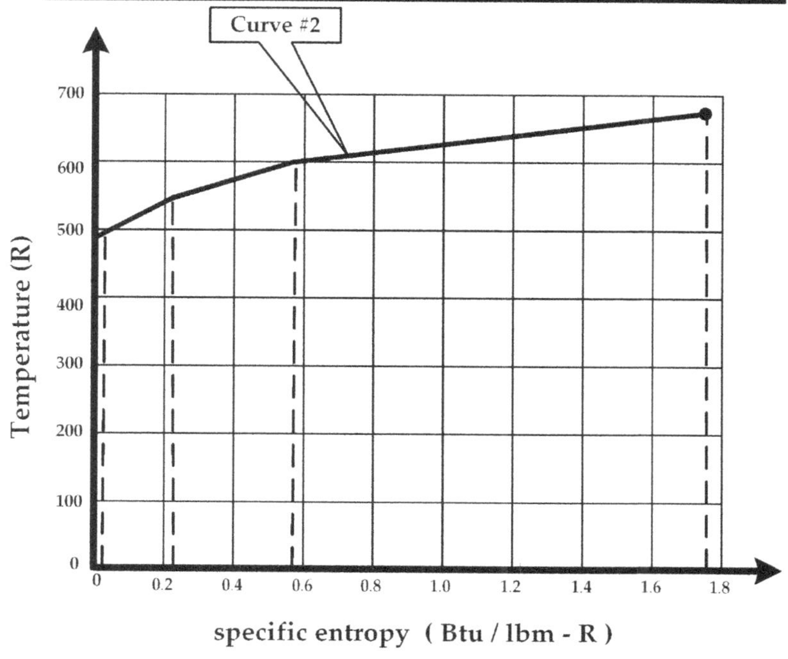

The Helmholtz function "F = U –T S" is defined as the
Free energy = Internal energy – (Temperature x Entropy)

Or the equation can be changed to define the internal energy "U = F + T S"
Internal energy = Free energy + (Temperature x Entropy)

Redefining the equation "U = F + T S" to "U = Uw + Uq" obtains the terms "internal energy work" and "internal energy heat".

From the last section we defined "Internal energy" as the area under a constant volume curve on a T-S diagram. The terms "internal energy work" and "internal energy heat" are the available and unavailable internal energy of a medium.

Internal energy = internal energy work + internal energy heat

Internal energy work is defined as the part of internal energy that is available to do work. The symbol for internal energy work is "Uw" and units of "Btu". The symbol for specific internal energy work is "uw" and units of "Btu/lbm."

Internal energy heat is the part of internal energy unavailable to do work The symbol for internal energy heat is "Uq" and with the units of "Btu". The symbol for specific internal energy heat is "uq" and with the units of "Btu/lbm".

Internal Energy....U = Uw + Uq = mu (Btu)

Specific internal energy... u = uw + uq (Btu/lbm

From the graph the internal energy is equal to the total area under the constant volume curve.

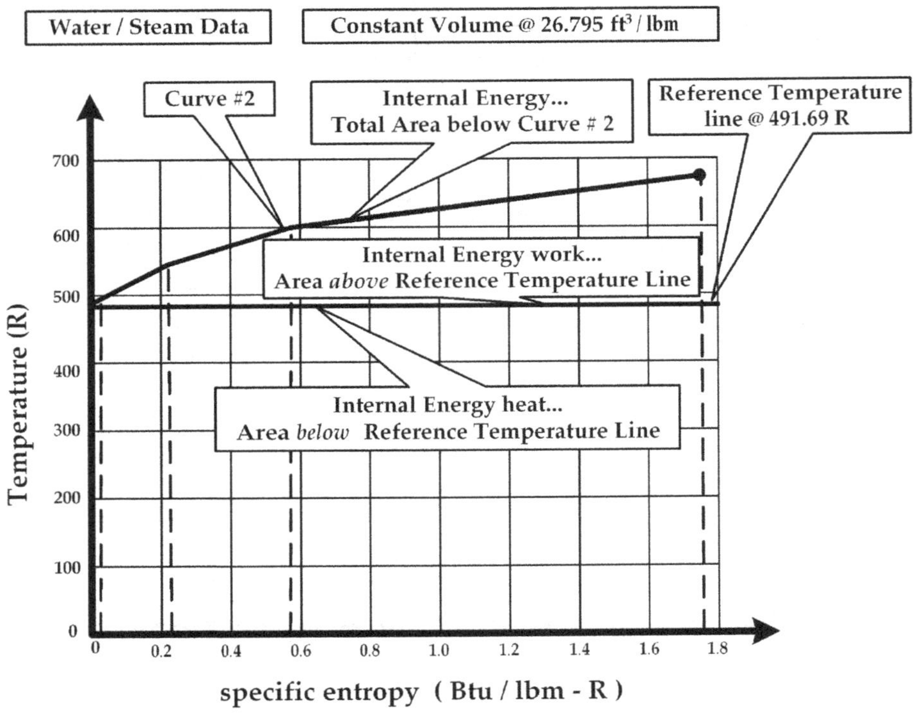

Next divided the internal energy (area under curve) into two sections

1. The "internal energy heat" is the area of under the curve below the temperature 491.68 °R. Internal energy Heat is simply Entropy times Reference Temperature

2. The "internal energy work" is the remaining area under the curve or simply: internal energy work is equal to internal energy (total area) – internal energy heat (area below 491.68 °R)

Looking at the graph above the calculation for the internal energy heat is simply entropy times Reference Temperature. Again the reference temperature is 491.68 °R. From Table below calculate uq and uw using the following equation:

1. Internal energy…"u" supplied by table
2. Internal energy heat "uq = Tr x s" = Reference Temperate time entropy
3. Internal energy work "uw = u – uq" = internal energy minus internal energy heat

Constant Volume @ 26.795 ft3 / lbm   Reference Temperature = 491.69 R

Temp	Pressure	Volume	Entropy	Int. Energy	uq	uw	Enthalpy	Pv work
°R	psia	ft³/lbm	Btu/lbm-°R	Btu/lbm	Btu/lbm	Btu/lbm	Btu/lbm	Btu/lbm
491.69	0	26.795	0.0000	0.00	uq0	uw 0	0.00	0.00
500	0.12331	26.795	0.0406	19.64	uq1	uw1	20.25	0.61
550	0.70631	26.795	0.2220	115.25	uq2	uw2	118.76	3.50
600	2.9176	26.795	0.5708	316.82	uq3	uw3	331.30	14.46
671.64	14.7	26.795	1.7577	1078.10	uq4	uw4	1151.00	72.87

Calculating: uq = specific internal energy heat   uw = specific internal energy work

Δ uq = (Tr) x Δ s = (Reference Temperature) x Δ specific entropy

Δ uw = Δ u - Δ uq = Δ specific internal energy - Δ specific internal energy heat

Using a starting reference temperature of 491.69 R and a "0" entropy the equations become:

uq = (Tr) x (s)

uw = (u) - (uq)

uq0 = (491.69) x (0) = 0 (Btu / lbm)

uw0 = (0) - (0) = 0 (Btu / lbm)

uq1 = (491.69) x (0.0406) = 19.99 (Btu / lbm)

uw1 = (19.64) - (19.99) = -0.34 (Btu / lbm)

Note: negative value of uw and hw near the reference temperature is an error in u, uw, uq or s value

uq2 = (491.69) x (0.2220) = 109.15 (Btu / lbm)

uw2 = (115.25) - (109.15) = 6.10 (Btu / lbm)

uq3 = (491.69) x (0.5708) = 280.63 (Btu / lbm)

uw3 = (316.82) - (280.63) = 36.19 (Btu / lbm)

uq4 = (491.69) x (1.7577) = 864.24 (Btu / lbm)

uw4 = (1078.10) - (864.24) = 213.86 (Btu / lbm)

Results from calculations:

Temp	Pressure	Volume	Entropy	Int. Energy	uq	uw	Enthalpy	Pv work
°R	psia	ft³/lbm	Btu/lbm-°R	Btu/lbm	Btu/lbm	Btu/lbm	Btu/lbm	Btu/lbm
491.69	0	26.795	0.0000	0.00	0.00	0.00	0.00	0.00
500	0.12331	26.795	0.0406	19.64	19.99	-0.34	20.25	0.61
550	0.70631	26.795	0.2220	115.25	109.15	6.10	118.76	3.50
600	2.9176	26.795	0.5708	316.82	280.63	36.19	331.30	14.46
671.64	14.7	26.795	1.7577	1078.10	864.24	213.86	1151.00	72.87

Assume the entropy data was not in hand, how would uq and uw be calculated ?

Constant Volume @ 26.795 ft3 / lbm **Reference Temperature = 491.69 R**

Temp	Pressure	Volume	Entropy	Int. Energy	uq	uw	Enthalpy	Pv work
°R	psia	ft³/lbm	Btu/lbm-°R	Btu/lbm	Btu/lbm	Btu/lbm	Btu/lbm	Btu/lbm
491.69	0	26.795	[redacted]	0.00	uq0	uw 0	0.00	0.00
500	0.12331	26.795	[redacted]	19.64	uq1	uw1	20.25	0.61
550	0.70631	26.795	[redacted]	115.25	uq2	uw2	118.76	3.50
600	2.9176	26.795	[redacted]	316.82	uq3	uw3	331.30	14.46
671.64	14.7	26.795	[redacted]	1078.10	uq4	uw4	1151.00	72.87

Calculating: uq = specific enthalpy heat
uw = specific enthalpy work

" Without entropy data "

Δ uq = (Tr) x Δ s = (Reference Temperature) x Δ specific entropy

Δ uw = Δ u - Δ uq = Δ specific internal energy - Δ specific internal energy heat

Replacing: Δ s for $\dfrac{\Delta u}{(Tm)} = \dfrac{\Delta \text{ specific internal energy}}{(\text{Mean Temperature})}$

$$\Delta uq = (Tr) \times \frac{\Delta u}{(Tm)} \qquad \Delta uw = \Delta u - \Delta uq$$

<u>Assuming data is linear between points :</u>

uq0 = 0 (Btu / lbm) uw0 = 0 (Btu / lbm)

$$uq1 - uq0 = (491.69) \times \frac{(19.64 - 0)}{(0.5)(500 + 491.69)} = 19.48 \text{ (Btu / lbm)}$$

uw1 - uw0 = (19.64 - 0) - (19.48) = 0.11 (Btu / lbm)

$$uq2 - uq1 = (491.69) \times \frac{(115.25 - 19.64)}{(0.5)(550 + 500)} = 89.54 \text{ (Btu / lbm)}$$

uw2 - uw1 = (115.25 - 19.64) - (89.54) = 6.07 (Btu / lbm)

$$uq3 - uq2 = (491.69) \times \frac{(316.82 - 115.25)}{(0.5)(550 + 600)} = 172.37 \text{ (Btu / lbm)}$$

uw3 - uw2 = (316.82 - 115.25) - (172.37) = 29.2 (Btu / lbm)

$$uq4 - uq3 = (491.69) \times \frac{(1078.1 - 316.82)}{(0.5)(600 + 671.64)} = 588.71 \text{ (Btu / lbm)}$$

uw4 - uw3 = (1078.1 -316.82) - (588.71) = 172.57 (Btu / lbm)

Constant Volume @ 26.795 ft3 / lbm

Reference Temperature = 491.69 R

Calculating:

uq = specific internal energy heat

uq = specific internal energy work

uq (n) = uq (n-1) + Δ uq

uw (n) = uw (n-1) + Δ uw

uq0 = 0 (Btu / lbm)

uw0 = 0 (Btu / lbm)

uq1 = uq0 + (uq1- uq0) = (0) + (19.48) = 19.48 (Btu / lbm)

uw1 = uw0 + (uw1- uw0) = (0) + (0.11) = 0.11 (Btu / lbm)

uq2 = uq1 + (uq2- uq1) = (19.48) + (89.54) = 109.02 (Btu / lbm)

uw2 = uw1 + (uw2- uw1) = (0.11) + (6.07) = 6.18 (Btu / lbm)

uq3 = uq2 + (uq3- uq2) = (109.02) + (172.37) = 281.39 (Btu / lbm)

uw3 = uw2 + (uw3- uw2) = (6.18) + (29.2) = 35.38 (Btu / lbm)

uq4 = uq3 + (uq4- uq3) = (281.39) + (588.71) = 870.10 (Btu / lbm)

uw4 = uw3 + (uw4- uw3) = (35.38) + (172.57) = 207.95 (Btu / lbm)

<u>Compare data with table below from previous calcations</u>

Temp	Pressure	Volume	Entropy	Int. Energy	uq	uw	Enthalpy	Pv work
°R	psia	ft³/lbm	Btu/lbm-°R	Btu/lbm	Btu/lbm	Btu/lbm	Btu/lbm	Btu/lbm
491.69	0	26.795	0.0000	0.00	0.00	0.00	0.00	0.00
500	0.12331	26.795	0.0406	19.64	19.99	-0.34	20.25	0.61
550	0.70631	26.795	0.2220	115.25	109.15	6.10	118.76	3.50
600	2.9176	26.795	0.5708	316.82	280.63	36.19	331.30	14.46
671.64	14.7	26.795	1.7577	1078.10	864.24	213.86	1151.00	72.87

Remember: the data is not linear between points, therefore the actual values of uw and uq are questionable

2.4 Summary… medium work and heat

Every point on a T-s diagram has two unique points… (P and v) or (T and s)

Enthalpy is the medium property that represents the thermal energy along a constant pressure path. (Curve #1…constant pressure @ 14.7 psia)

a. The "enthalpy heat" is the area of under the curve below the temperature 491.69 °R. Enthalpy Heat is simply Entropy times Reference Temperature

b. The "enthalpy work" is the remaining area under the curve or simply: enthalpy work is equal to enthalpy (total area) – enthalpy heat (area below 491.69 °R)

Internal Energy is the medium property that represents the thermal energy along a constant volume path. (Curve #2… constant volume @ 26.795 ft^3/lbm)

1. The "internal energy heat" is the area of under the curve below the temperature 491.69 °R. Internal energy Heat is simply Entropy times Reference Temperature

2. The "internal energy work" is the remaining area under the curve or simply: internal energy work is equal to internal energy (total area) – internal energy heat (area below 491.68 °R)

Pv work represents the thermal energy between the constant pressure path and the constant volume path. Also Pv work area is above the reference temperature 491.69 °R. Therefore

Pv work = Enthalpy - Internal Energy = Enthalpy work –Internal Energy work

Summary of data and graph.

Water / Steam Data				Constant Pressure @ 14.7 psia				
Temp	Pressure	Volume	Entropy	Int. Energy	Enthalpy	hq	hw	Pv work
°R	psia	ft³/lbm	Btu/lbm-°R	Btu/lbm	Btu/lbm	Btu/lbm	Btu/lbm	Btu/lbm
491.69	14.7	0	0.0000	0.00	0.00	0.00	0.00	0.00
500	14.7	0.016019	0.0169	8.37	8.41	8.30	0.11	0.04
600	14.7	0.016294	0.1992	108.38	108.42	97.95	10.47	0.04
671.64	14.7	0.016714	0.3124	180.25	180.30	153.59	26.71	0.05
671.64	14.7	26.795	1.7577	1078.10	1151.00	864.24	286.76	72.87

Water / Steam Data				Constant Volume @ 26.795 ft³ / lbm				
Temp	Pressure	Volume	Entropy	Int. Energy	uq	uw	Enthalpy	Pv work
°R	psia	ft³/lbm	Btu/lbm-°R	Btu/lbm	Btu/lbm	Btu/lbm	Btu/lbm	Btu/lbm
491.69	0	26.795	0.0000	0.00	0.00	0.00	0.00	0.00
500	0.12331	26.795	0.0406	19.64	19.99	-0.34	20.25	0.61
550	0.70631	26.795	0.2220	115.25	109.15	6.10	118.76	3.50
600	2.9176	26.795	0.5708	316.82	280.63	36.19	331.30	14.46
671.64	14.7	26.795	1.7577	1078.10	864.24	213.86	1151.00	72.87

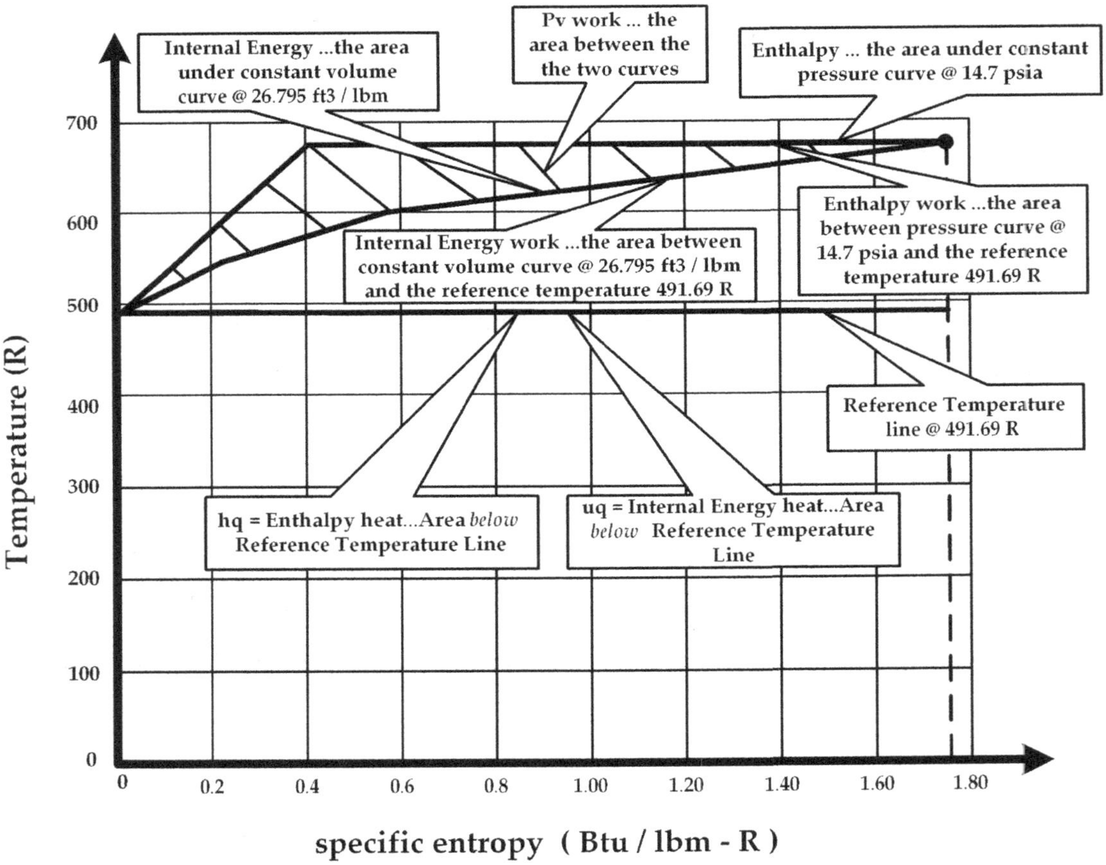

Below is a list an equation summary:

Enthalpy

$H = m \times h = U + (0.185)Pv = Hw + Hq$ (Btu)

Specific Enthalpy

$h = H / m$ (Btu/lbm)

$h = u + (0.185)Pv$ (Btu/lbm)

$h = hw + hq$ (Btu/lbm)

$h = T\,mh \times s$ (Btu/lbm)

$\Delta h = h2 - h1$ (Btu/lbm)

$\Delta h = (hw\,2 + hq\,2) - (hw\,1 + hq\,1)$ (Btu/lbm)

$\Delta h = (hw\,2 - hw\,1) + (hq\,2 - hq\,1)$ (Btu/lbm)

$\Delta h = (uw\,2 + uq\,2 + (0.185)Pv\,2) - (uw\,1 + uq1 + (0.185)Pv\,1)$ (Btu/lbm)

$\Delta h = (uw\,2 - uw\,1) - (uq\,2 - uq\,1) + ((0.185)Pv\,2 - (0.185)Pv\,1)$ (Btu/lbm)

$\Delta h = (T\,mh2 \times s2) - (Tmh1 \times s1)$ (Btu/lbm)

$\Delta h = \Delta Tmh \times \Delta s$ (Btu/lbm)

Specific Enthalpy Heat

$hq = Hq / m$ (Btu/lbm)

$hq = uq$ (Btu/lbm)

$hq = h - hw$ (Btu/lbm)

$hq = T\,r \times s$ (Btu/lbm)

$\Delta hq = (hq\,2 - hq\,1)$ (Btu/lbm)

$\Delta hq = (h\,2 - h\,1) - (hw\,2 - hw\,1)$ (Btu/lbm)

$\Delta hq = Tr \times (s\,2 - s\,1)$ (Btu/lbm)

$\Delta hq = Tr \times (\Delta s)$ (Btu/lbm)

Specific Enthalpy Work

$hw = Hw / m$ (Btu/lbm)

$hw = h - hq$ (Btu/lbm)

$hw = uw + (0.185)\,Pv$ (Btu/lbm)

$hw = (T\,mh - Tr) \times s$ (Btu/lbm)

$\Delta hw = (hw\,2 - hw\,1)$ (Btu/lbm)

$\Delta hw = (h\,2 - h\,1) - (hq\,2 - hq1)$ (Btu/lbm)

$\Delta hw = (uw\,2 + (0.185)Pv\,2) - (uw\,1 + (0.185)Pv\,1)$ (Btu/lbm)

$\Delta hw = ((T\,mh2 - Tr) \times s2) - ((T\,mh1 - Tr) \times s\,1)$ (Btu/lbm)

Calculate hw and hq for the example below:

Solve for hw and hq for the following data point...

Temp	Pressure	Volume	Entropy	Enthalpy
°R	psia	ft³/lbm	Btu/lbm-°R	Btu/lbm
1500	800	1.08	1.6976	1535.70

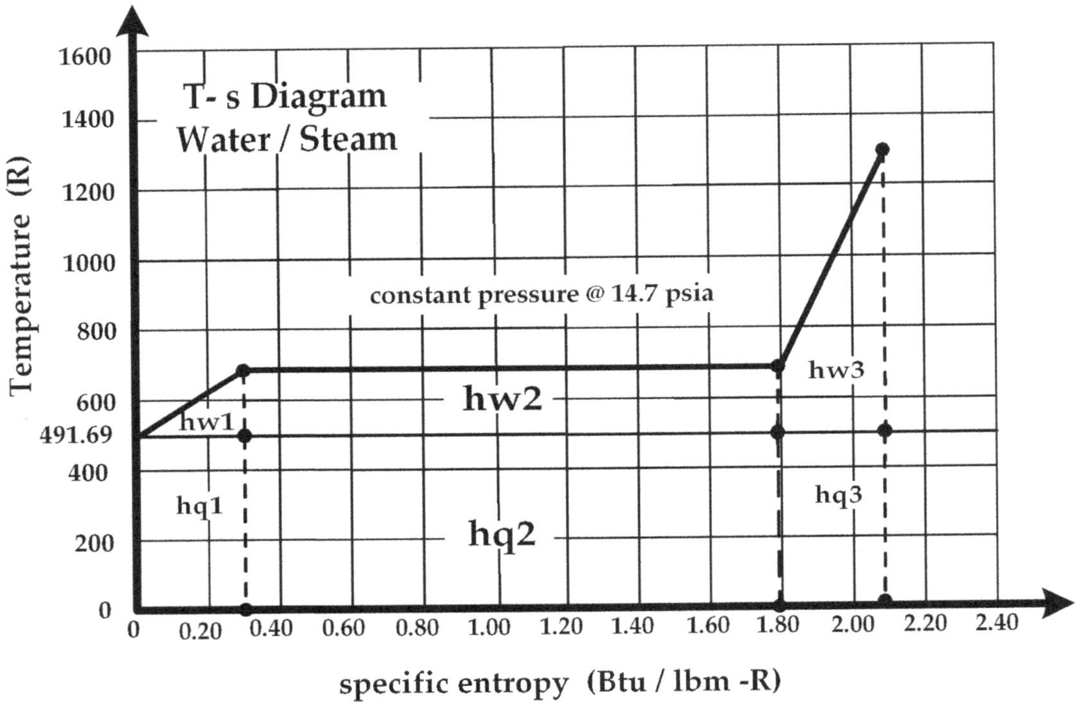

Constant Pressure @ 14.7 psia

Temp	Pressure	Volume	Entropy	Enthalpy
°R	psia	ft³/lbm	Btu/lbm-°R	Btu/lbm
491.69	800	0.02	0.0001	2.41
977.94	800	0.02	0.7117	510.24
977.94	800	0.57	1.4171	1200.10
1500	800	1.08	1.6976	1535.70

cont'... Constant Pressure @ 800 psia

<u>Calculating</u>: Δ specific Pv work = (0.185) (P) (v)

Pv work0 = (0.185) (P1) (v1) = 0 Btu/lbm

Pv work1 = (0.185) (P1) (v1 -v0) = (0.185) (800) (0.02 -0) = 2.96 Btu/lbm

Pv work2 = (0.185) (P2) (v2-v1) = (0.185) (800) (0.57-0.02) = 81.4 Btu/lbm

Pv work3 = (0.185) (P3) (v3-v2) = (0.185) (800) (1.08 - 0.57) = 75.48 Btu/lbm

<u>u = specific internal energy</u>

Δ specific internal energy = specific enthalpy - specific Pv work

Δu = Δh - Δ(0.185) (P) (v)

u1 = (h1 - h0)-[(0.185) (P) (v1 - v0)] = (510.24 - 0) - (2.96) = 507.28 Btu/lbm

u2 = (h2 - h1)-[(0.185) (P) (v2 - v1)] = (1200.10 - 510.24) - (81.4) = 608.46 Btu/lbm

u3 = (h3 - h2)-[(0.185) (P) (v3 - v2)] = (1535.7 - 1200.10) - (75.48) = 260.12 Btu/lbm

u = u1 + u2 + u3 = (507.28) + (608.46) + (260.12) = 1375.86 Btu/lbm

<u>uq = specific internal energy heat</u>

Δhq = Δuq

uq1 = hq1 = 349.94

uq2 = hq2 = 346.84

uq3 = hq3 = 137.92

uq = uq1 + uq2 + uq3

uq = (349.94) + (346.84) + (137.92)

uq = 834.7 Btu /lbm

<u>uw = specific internal energy work</u>

Δuw = Δu - Δuq = Δhw - Δ(0.185) (P) (v)

uw1 = u1 - uq1 = 507.28 - 349.94 = 157.34 Btu/lbm

uw2 = u2 - uq2 = 608.46 - 346.84 = 261.62 Btu/lbm

uw3 = u3 - uq3 = 260.12 - 137.92 = 122.2 Btu/lbm

uw = uw1 + uw2 + uw3

uw = (157.34) + (261.62) + (122.2)

uw = 541.16 Btu/lbm

u = uw + uq = 541.16 + 834.7 = 1375.86 Btu/lbm

Temp	Pressure	Volume	Entropy	Enthalpy	Int. Energy	uq/hq	hw	uw	Pv work	Tmh
°R	psia	ft³/lbm	Btu/lbm-°R	Btu/lbm	Btu/lbm	Btu/lbm	Btu/lbm	Btu/lbm	Btu/lbm	°R
491.69	800	0.02	0.0001	2.41	2.41	0.04	2.37	2.37	0.00	491.69
977.94	800	0.02	0.7117	510.24	507.15	349.95	160.29	157.20	3.09	716.90
977.94	800	0.57	1.4171	1200.10	1115.86	696.77	503.33	419.09	84.24	846.87
1500	800	1.08	1.6976	1535.70	1375.62	834.69	701.01	540.93	160.08	904.63

Here is a different way to look at energy using a T-s graph with mass as the z-axis:

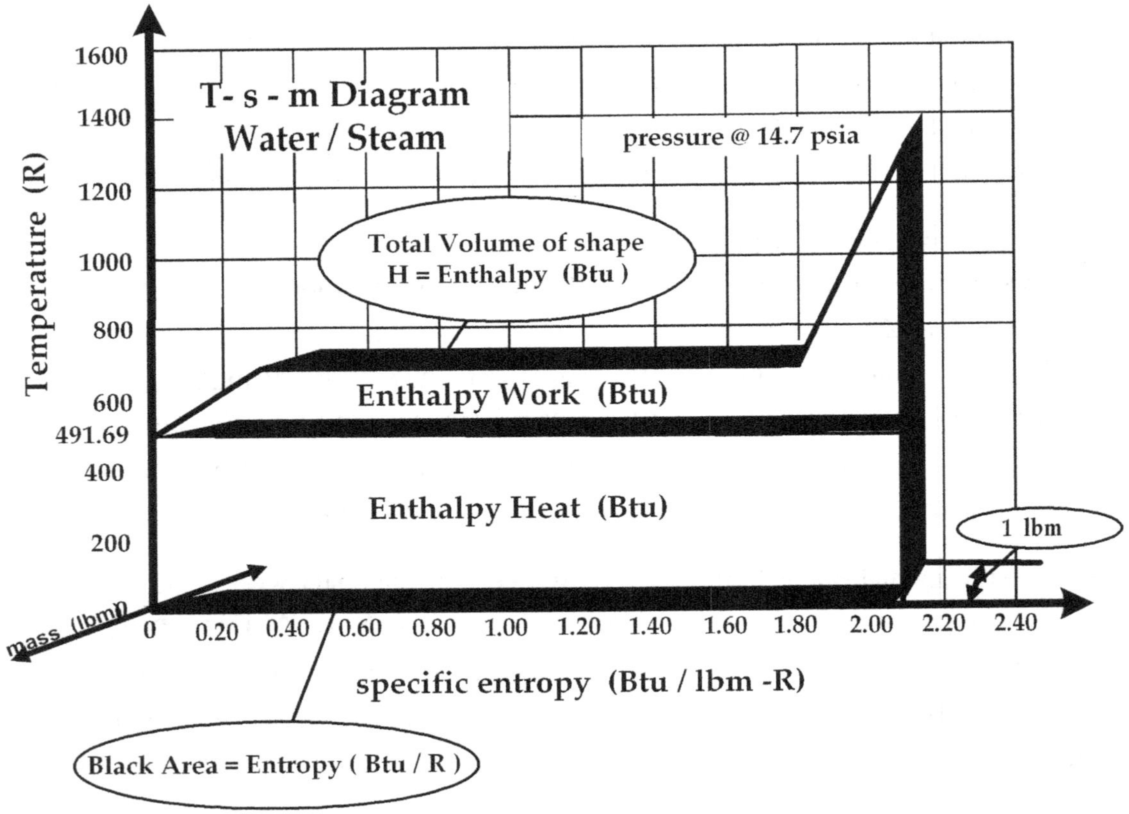

Image using "time" as the forth dimension. If the volume from the graph above represented one minute of time then image stacking each volume along a time stamp axis in a one minute intervals. The four axes…Temperature, specific entropy, mass and time.

3 Processes Heat and Work entering / leaving the medium

"The Carnot cycle limits our thinking"
Fred J. Weber

3.1 What is a Process ?

A process is a change in the medium's property value at a particular time. This chapter is a glimpse of the various processes in thermodynamics. The data describing each process is from a typical water / steam table. This medium was selected because of the many applications in industry that operate between all three phases (liquid, liquid- vapor and vapor phase). The reference temperature 491.69 °R will be used in the calculation enthalpy work, enthalpy heat, internal energy work and internal energy heat. Three items will be used in the discussion of each process:

1. Data about the medium at the constant variable such as constant pressure, entropy or enthalpy.

Temp	Pressure	Volume	Entropy	Enthalpy	uq / hq	hw	Pv work	Tmh
°R	psia	ft³/lbm	Btu/lbm-°R	Btu/lbm	Btu/lbm	Btu/lbm	Btu/lbm	°R
491.69	2400	0.016	0.0002	7.2	0.1	7.1	7.1	491.69
500	2400	0.016	0.0167	15.4	8.2	7.2	7.1	919.36
700	2400	0.017	0.3508	214.0	172.5	41.5	7.5	610.13
900	2400	0.019	0.6107	421.2	300.3	120.9	8.4	689.71
1100	2400	0.025	0.8638	675.3	424.7	250.5	11.3	781.71
1121.8	2400	0.028	0.9033	719.2	444.2	275.0	12.4	796.12
1121.8	2400	0.141	1.2452	1102.6	612.3	490.3	62.5	885.48
1300	2400	0.261	1.4490	1345.0	712.5	632.5	115.9	928.23
1500	2400	0.336	1.5526	1489.4	763.4	726.0	149.0	959.29

2. T-s diagram (Temperature – specific entropy). To understand any process it is important to locate the initial and final data point for a process and the path the process took.

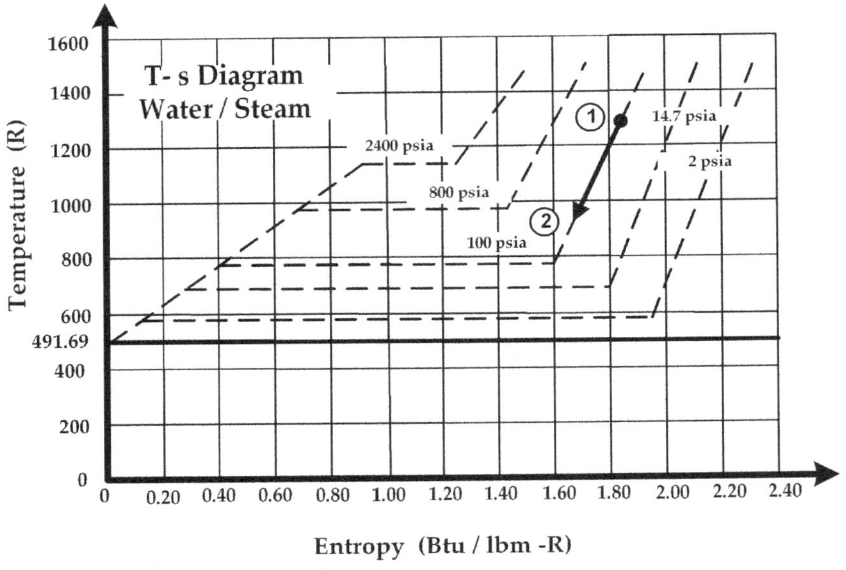

3. Thermo-diagram that shows the changes in properties values. Imagine taking a picture of the gauge and instrument readings on a piece of equipment at a particular time. Then wait a hour and take the same pictures of gauges. Understanding the changes of a medium before and after a process defines the change in energy, the change in energy available to do work (hw) and the change in energy unavailable to do work (hq).

Weber's Thermo Diagram

H2O	Temp	Pressure	Volume	Entropy	Enthalpy	uq / hq	hw	Pv work	Tmh
491.69	°R	psia	ft³/lbm	Btu/lbm-°R	Btu/lbm	Btu/lbm	Btu/lbm	Btu/lbm	°R
1- Initial	700	2400	0.017	0.3508	214.0	172.5	41.5	7.5	610.13
2- Final	1500	2400	0.336	1.5526	1489.4	763.4	726.0	149.0	959.29
(2 - 1)	800.0	0.0	0.32	1.2018	1275.4	590.9	684.5	141.5	349.2
1	lbm /hr		0.32	1.2018	1275.4	590.9	684.5	141.5	

And within the thermo-diagram we need to know the medium, mass and time.

The first factor we need to identify is the medium. Compare the thermodynamic properties of water/ steam vs. ammonia at 1 lbm @ 600 (°R) and 14.7 psia.

Media

Reference Temperature

H2O	Temp	Pressure	Volume	Entropy	Enthalpy	uq / hq	hw	Pv work	Tmh
491.69	°R	psia	ft³/lbm	Btu/lbm-°R	Btu/lbm	Btu/lbm	Btu/lbm	Btu/lbm	°R
1- Initial	600	14.7	0.016	0.1992	108.4	98.0	10.5	0.0	544.22

NH3	Temp	Pressure	Volume	Entropy	Enthalpy	uq / hq	hw	Pv work	Tmh
351.9	°R	psia	ft³/lbm	Btu/lbm-°R	Btu/lbm	Btu/lbm	Btu/lbm	Btu/lbm	°R
1- Initial	600	14.7	25.545	1.7484	759.8	615.3	144.6	69.5	434.59

Display the medium's reference temperature, the temperature when the entropy and enthalpy are zero. Next is to display the mass of the medium and the time frame the process occurred. The time frame is a specific time (one minute, one hour or one day) between the initial and final readings. For example a non flow process may have a mass of 5 lbm with a time frame of one minute or a flow process at 5 lbm per minute. The time frame is required for a "system" where the transfer of energy of both processes need to be in the same time frame.

This section contains several processes of the medium (water) while holding one of the medium's properties constant. In the real world a piece of equipment typical doesn't follow at constant variable like enthalpy, pressure or entropy but a combination of processes. The key is to understand what processes are causing the deviations. Example of this is the turbine. An ideal turbine typical is defined as a constant entropy process but in the real world the turbine could be a combination of processes:

Constant specific entropy and constant pressure.

Constant specific entropy and constant specific enthalpy.

Constant specific enthalpy work, and constant specific entropy and constant specific enthalpy.

The answer will come from a closer analysis of the turbine. For now we will be concerned with the constant property process (i.e. constant pressure, volume and etc.) and a combination of processes to achieve the same.

While reviewing this section identify which process is a source, sink or neither.

A source is a process that releases energy ($-\Delta H$) and work ($-\Delta Hw$)

A sink on the other hand is a process that acts like a sponge and absorbs energy. The sink soaks up the energy ($+\Delta H$), the work ($+\Delta Hw$), the heat ($+\Delta Hq$) and converts any left over work to heat ($+\Delta Hq$). There are more restrictions that define a source and sink but for now simply identify the change in energy, work and heat of each process.

.

3.2 Constant Pressure...P1 = P2 (psia)

The thermodynamic process frequently applied in industry today is the Constant Pressure process. Every time you see a heat exchanger you are typically looking at a constant pressure process application. Examples of this are a car radiator, air conditioner's evaporator or a power plant's condenser. Besides defining the relationship between temperature, enthalpy and entropy as discussed earlier, it is the cornerstone in the study of thermodynamics.

To begin the discussion let's look at the constant pressure data of steam/water.

Temp °R	Pressure psia	Volume ft³/lbm	Entropy Btu/lbm-°R	Enthalpy Btu/lbm	uq / hq Btu/lbm	hw Btu/lbm	Pv work Btu/lbm	Tmh °R
491.69	2400	0.016	0.0002	7.2	0.1	7.1	7.1	491.69
500	2400	0.016	0.0167	15.4	8.2	7.2	7.1	919.36
700	2400	0.017	0.3508	214.0	172.5	41.5	7.5	610.13
900	2400	0.019	0.6107	421.2	300.3	120.9	8.4	689.71
1100	2400	0.025	0.8638	675.3	424.7	250.5	11.3	781.71
1121.8	2400	0.028	0.9033	719.2	444.2	275.0	12.4	796.12
1121.8	2400	0.141	1.2452	1102.6	612.3	490.3	62.5	885.48
1300	2400	0.261	1.4490	1345.0	712.5	632.5	115.9	928.23
1500	2400	0.336	1.5526	1489.4	763.4	726.0	149.0	959.29

Observations

1. Pressure is constant
2. Properties have an increasing value... T, p, v, s, and h
3. Two readings at the same temperature of 1121.8 degrees R which represents the liquid to superheat phase of the medium (latent heat).
4. The data starts from the reference temperature 491.69 degrees R. This allows the user to see the change in property values from liquid, liquid-vapor phase to superheat state

Lets look at constant pressure process as energy is added (+).

H2O	Temp	Pressure	Volume	Entropy	Enthalpy	uq / hq	hw	Pv work	Tmh
491.69	°R	psia	ft³/lbm	Btu/lbm-°R	Btu/lbm	Btu/lbm	Btu/lbm	Btu/lbm	°R
1- Initial	700	2400	0.017	0.3508	214.0	172.5	41.5	7.5	610.13
2- Final	1500	2400	0.336	1.5526	1489.4	763.4	726.0	149.0	959.29
(2 - 1)	800.0	0.0	0.32	1.2018	1275.4	590.9	684.5	141.5	349.2
1	lbm /hr		0.32	1.2018	1275.4	590.9	684.5	141.5	

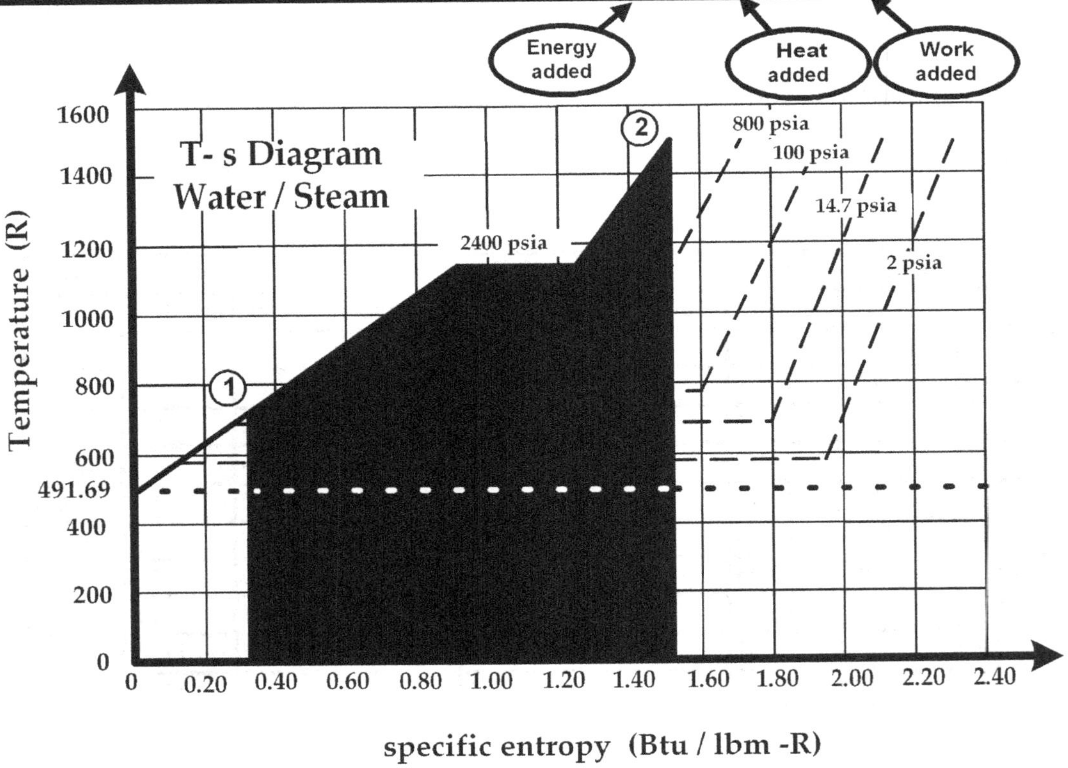

Observations

1. Pressure constant @ 2400 psia

2. Energy added to media = (black area) x (mass)
black area = change in specific enthalpy (h)

3. Work added to media = (black area above 491.69 R) x (mass)
black area above 491.69 R = change in specific enthalpy work (hw)

4. Heat added to media = (black area below 491.69 R) x (mass)
black area below 491.69 R = change in specific enthalpy heat (hq)

5. Change in specific internal energy (not shown)= h - Pv work = 1133.9 btu / lbm

Here is an example of another constant pressure process with energy removed (-) from the medium.

H2O	Temp	Pressure	Volume	Entropy	Enthalpy	uq / hq	hw	Pv work	Tmh
491.69	°R	psia	ft³/lbm	Btu/lbm-°R	Btu/lbm	Btu/lbm	Btu/lbm	Btu/lbm	°R
1- Initial	1300	100	7.691	1.8626	1451.2	915.8	535.4	142.3	779.13
2- Final	900	100	5.203	1.6772	1249.7	824.7	425.0	96.3	745.11
(2 - 1)	-400.0	0.0	-2.49	-0.1854	-201.5	-91.2	-110.3	-46.0	-34.0
1	lbm /hr		-2.49	-0.1854	-201.5	-91.2	-110.3	-46.0	

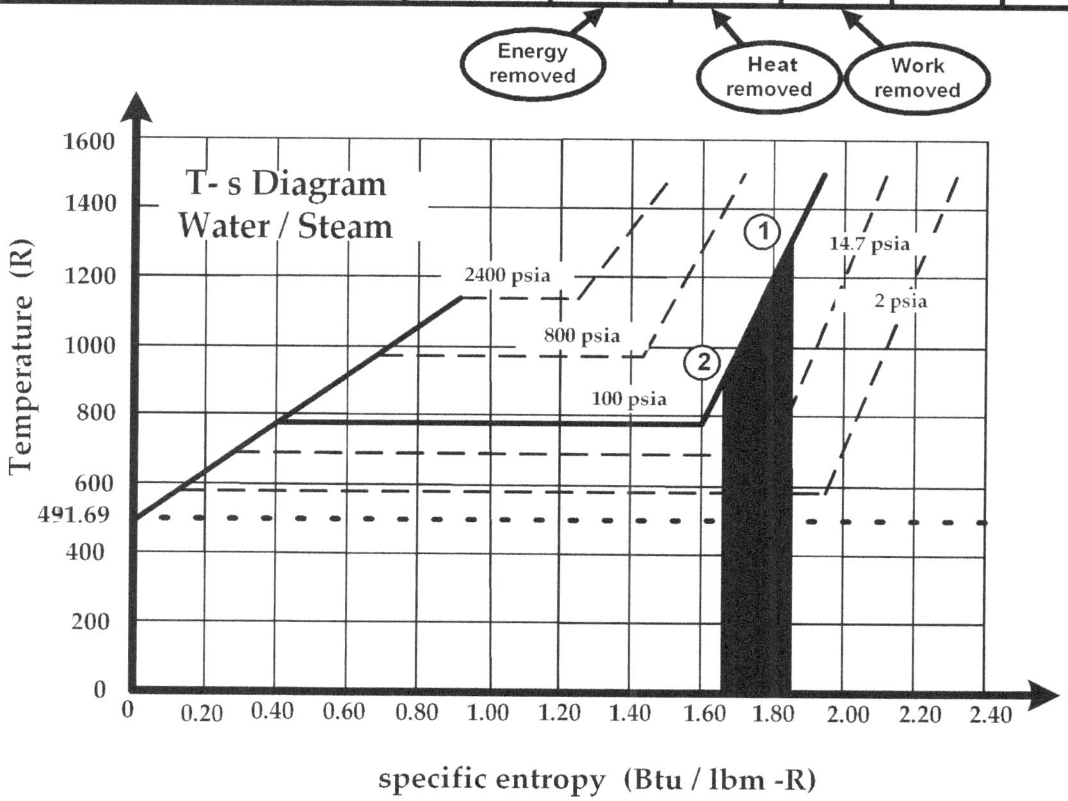

Observations

1. Pressure constant @ 100 psia

2. Energy removed from media = (black area) x (mass)
black area = change in specific enthalpy (h)

3. Work removed from media = (black area above 491.69 R) x (mass)
black area above 491.69 R = change in specific enthalpy work(hw)

4. Heat removed from media = (black area below 491.69 R) x (mass)
black area below 491.69 R = change in specific enthapy heat (hq)

5. Change in specific internal energy (not shown)= h - Pv work = - 155.5 Btu/lbm

In the previous example we assumed a constant pressure process based on the initial and final pressure were the same. But what if the path of the medium did not go though a heat exchanger but went though two unknown devices that gave the appearance of a constant pressure process? Below is an example of the medium moving through two different devices. One process is at constant entropy and the second one is at a constant volume.

H2O	Temp	Pressure	Volume	Entropy	Enthalpy	uq / hq	hw	Pv work	Tmh
491.69	°R	psia	ft³/lbm	Btu/lbm-°R	Btu/lbm	Btu/lbm	Btu/lbm	Btu/lbm	°R
1- Initial	1300	100	7.691	1.8626	1451.2	915.8	535.4	142.3	779.13
2- Final	1450	164.78	5.203	1.8626	1526.4	915.8	610.6	158.6	819.50
(2 - 1)	149.8	64.8	-2.49	0.0000	75.2	0.0	75.2	16.3	40.4
1	lbm /hr		-2.49	0.0000	75.2	0.0	75.2	16.3	

H2O	Temp	Pressure	Volume	Entropy	Enthalpy	uq / hq	hw	Pv work	Tmh
491.69	°R	psia	ft³/lbm	Btu/lbm-°R	Btu/lbm	Btu/lbm	Btu/lbm	Btu/lbm	°R
1- Initial	1450	164.78	5.203	1.8626	1526.4	915.8	610.6	158.6	819.50
2- Final	900	100	5.203	1.6772	1249.7	824.7	425.0	96.3	745.11
(2 - 1)	-549.8	-64.8	0.00	-0.1854	-276.7	-91.2	-185.5	-62.4	-74.4
1	lbm /hr		0.00	-0.1854	-276.7	-91.2	-185.5	-62.4	

1	lbm /hr		-2.49	-0.1854	-201.5	-91.2	-110.3	-46.0	

Results of combining both processes same as the previous example

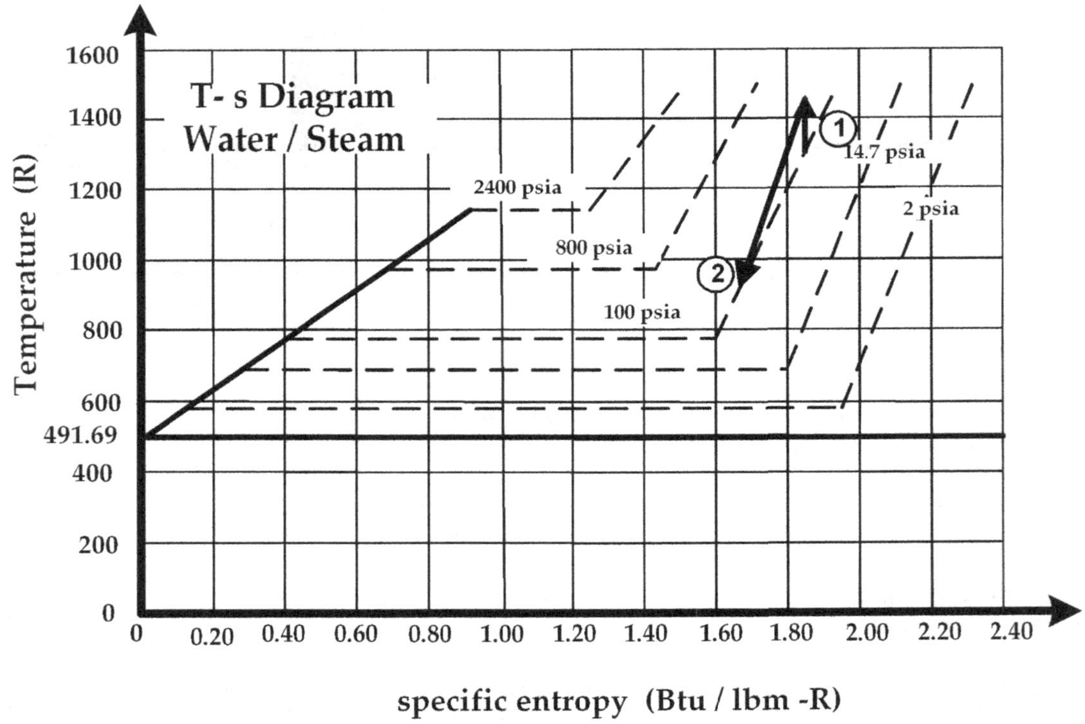

3.3 Constant Specific Volume...v1 = v2

This Constant Volume process is another well known thermodynamic process applied in industry. Every time you see a reservoir of a medium you are typically looking at a constant volume process application. Also remember the important contribution the constant volume process made in defining the relationship between temperature, internal energy and entropy. As discussed earlier, it is the second cornerstone in the study of thermodynamics.

To begin the discussion let's look at the constant volume data of steam/water.

Temp	Pressure	Volume	Entropy	Enthalpy	uq / hq	hw	Pv work	Tmh
°R	psia	ft³/lbm	Btu/lbm-°R	Btu/lbm	Btu/lbm	Btu/lbm	Btu/lbm	°R
585.7	2	0.328	0.1782	95.9	87.6	8.3	0.1	538.15
671.6	14.7	0.328	0.3292	191.6	161.9	29.7	0.9	581.99
787.5	100	0.328	0.5540	361.2	272.4	88.8	6.1	652.05
977.9	800	0.328	1.1069	896.7	544.3	352.4	48.5	810.06
1477.6	2400	0.328	1.5427	1474.6	758.5	716.1	145.6	955.86

Observations:

1. Volume is constant
2. Properties have an increasing value... T, p, v, s, and h
4. It is difficult to determine the liquid, liquid-vapor and superheat phases of the medium

Here is an example of a constant specific volume process with energy removed (-) from the medium. Notice the work removed from a constant volume process.

H2O	Temp	Pressure	Volume	Entropy	Enthalpy	uq / hq	hw	Pv work	Tmh	Int Egy	uw
491.69	°R	psia	ft³/lbm	Btu/lbm-°R	Btu/lbm	Btu/lbm	Btu/lbm	Btu/lbm	°R	Btu/lbm	Btu/lbm
1- Initial	977.9	800	0.328	1.1069	896.7	544.3	352.4	48.5	810.06	848.1	303.9
2- Final	787.5	100	0.328	0.5540	361.2	272.4	88.8	6.1	652.05	355.2	82.8
(2 - 1)	-190.5	-700.0	0.000	-0.5529	-535.4	-271.9	-263.6	-42.5	-158.01	-493.0	-221.1
1	lbm /hr		0.000	-0.5529	-535.4	-271.9	-263.6	-42.5		-493.0	-221.1

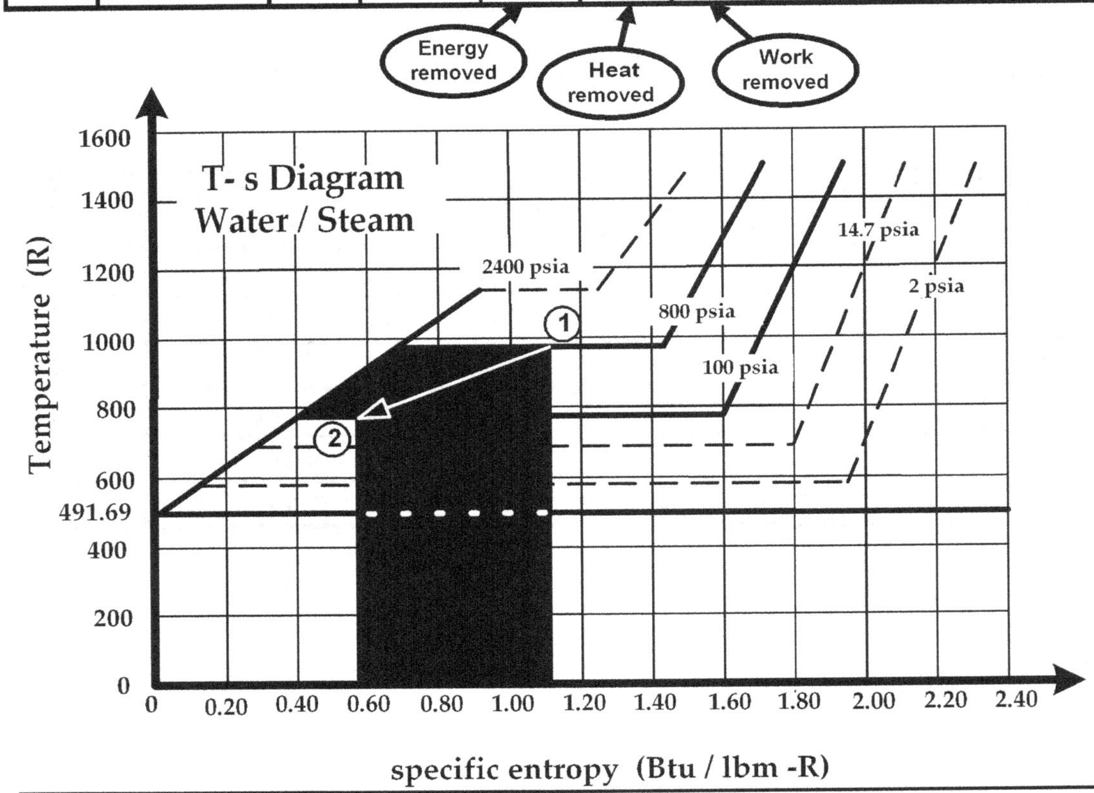

Observations

1. Specific volume constant @ 0.328 ft^3 / lbm

2. Energy removed from media = (black area) x (mass)
black area = change in specific enthalpy (h)

3. Work removed from media = (black area above 491.69 R) x (mass)
black area above 491.69 R = change in specific enthalpy work (hw)

4. Heat removed from media = (black area below 491.69 R) x (mass)
black area below 491.69 R = change in specific enthalpy heat (hq)

5. Change in specific internal energy (u) is the black area below the (1)-(2) line

6. Change in specific internal energy work (uw) is the
black area below the (1) - (2) line and above the 491.69 R line

7. Change in specific internal energy heat (uq) is the black area below the 491.69 R line and is the same as specific enthalpy heat (hq)

Here is ar example of another constant specific volume process with energy added (+) to the medium.

H2O	Temp	Pressure	Volume	Entropy	Enthalpy	uq / hq	hw	Pv work	Tmh
491.69	°R	psia	ft³/lbm	Btu/lbm-°R	Btu/lbm	Btu/lbm	Btu/lbm	Btu/lbm	°R
1- Initial	977.9	800	0.328	1.1069	896.7	544.3	352.4	48.5	810.06
2- Final	1477.6	2400	0.328	1.5427	1474.6	758.5	716.1	145.6	955.86
(2 - 1)	499.7	1600.0	0.000	0.4358	577.9	214.3	363.7	97.1	145.79
1	lbm /hr		0.000	0.4358	577.9	214.3	363.7	97.1	

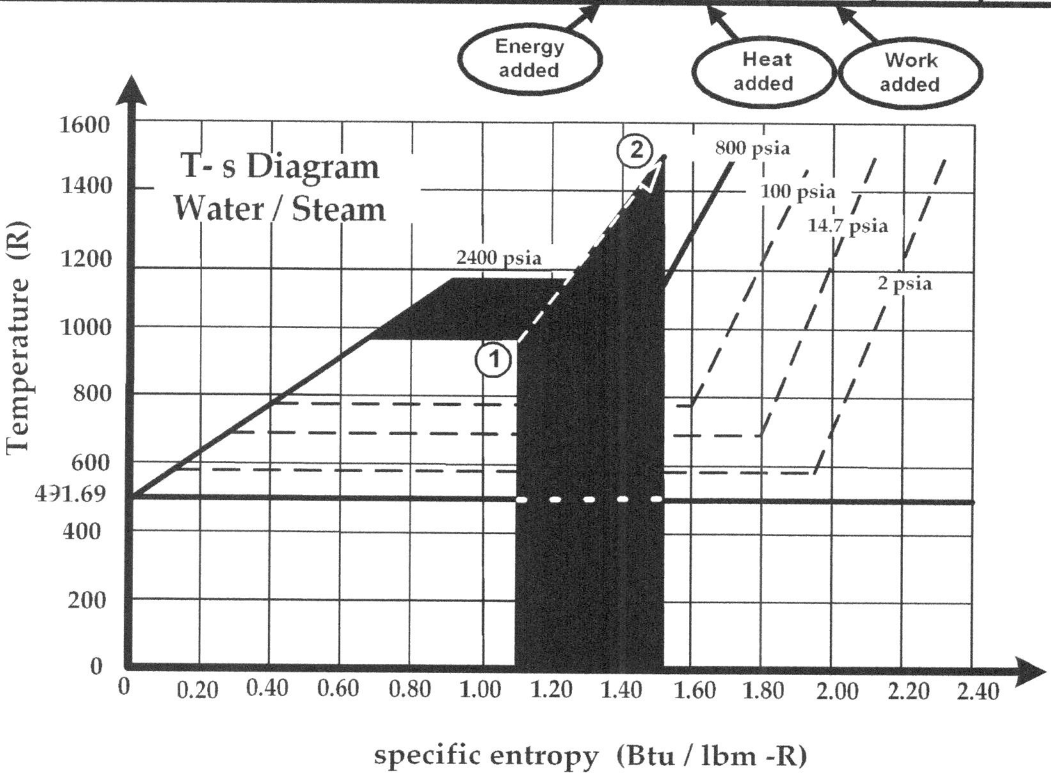

Observations

1. Specific volume constant @ 0.328 ft^3 / lbm

2. Energy added to media = (black area) x (mass)
black area = change in specific enthalpy (h)

3. Work added to media = (black area above 491.69 R) x (mass)
black area above 491.69 R = change in specific enthapy work (hw)

4. Heat added to media = (black area below 491.69 R) x (mass)
black area below 491.69 R = change in specific enthapy heat (hq)

5. Change in specific internal energy is the black area below the (1)-(2) line
u = = h - Pv work = 577.9 - 97.1 = 480.8 btu / lbm

Below is the medium from the previous constant volume process moving through two different processes to achieve the same final results. One process is at constant entropy and the second one is at a constant pressure.

H2O	Temp	Pressure	Volume	Entropy	Enthalpy	uq / hq	hw	Pv work	Tmh
491.69	°R	psia	ft³/lbm	Btu/lbm-°R	Btu/lbm	Btu/lbm	Btu/lbm	Btu/lbm	°R
1- Initial	977.9	800	0.328	1.1069	896.7	544.3	352.4	48.5	810.06
2- Final	1121.8	2400	0.095	1.1069	947.5	544.3	403.3	42.2	856.02
(2 - 1)	143.9	1600.0	-0.233	0.0000	50.9	0.0	50.9	-6.3	45.96
1	lbm /hr		-0.233	0.0000	50.9	0.0	50.9	-6.3	

H2O	Temp	Pressure	Volume	Entropy	Enthalpy	uq / hq	hw	Pv work	Tmh
491.69	°R	psia	ft³/lbm	Btu/lbm-°R	Btu/lbm	Btu/lbm	Btu/lbm	Btu/lbm	°R
1- Initial	1121.8	2400	0.095	1.1069	947.5	544.3	403.3	42.2	856.02
2- Final	1477.6	2400	0.328	1.5427	1474.6	758.5	716.1	145.6	955.86
(2 - 1)	355.8	0.0	0.233	0.4358	527.1	214.3	312.8	103.4	99.84
1	lbm /hr		0.233	0.4358	527.1	214.3	312.8	103.4	

Combined Results

1	lbm /hr		0.000	0.4358	577.9	214.3	363.7	97.1	

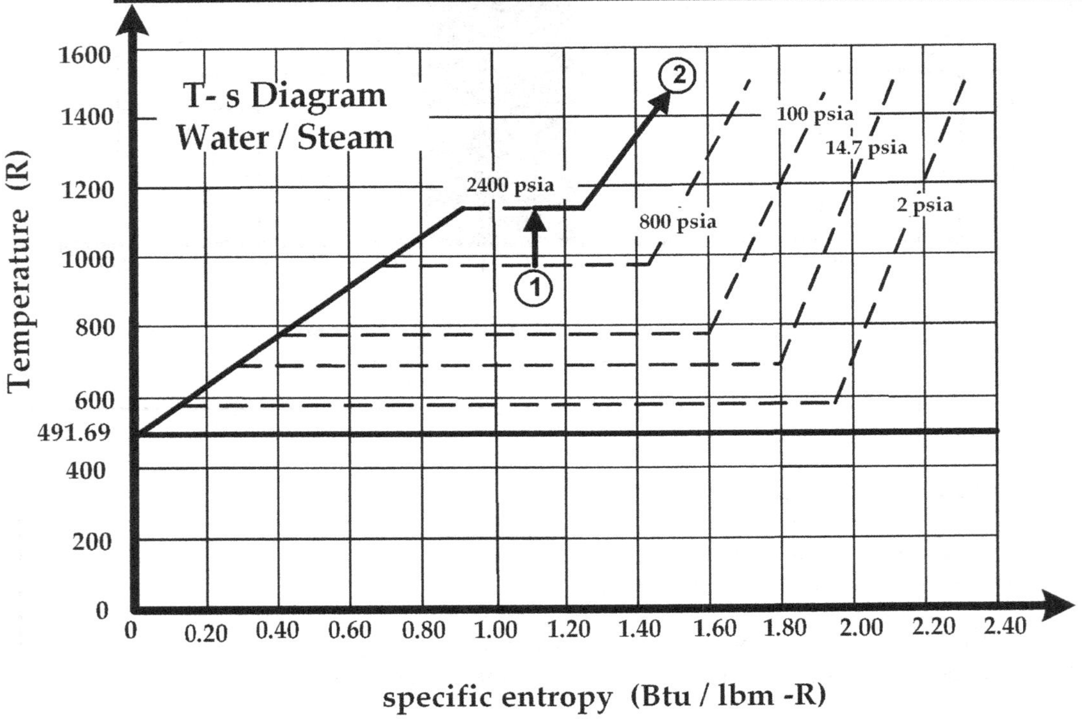

3.4 Constant Specific Entropy...s1 = s2

The title for this section should have been labeled:

3.4 Constant Specific Entropy s1 = s2 , Constant Specific enthalpy heat hq1 = hq2 and Constant Specific internal energy heat uq1 = uq2

But there was a space limitation so it is condensed.

This powerful process only adds or removes work energy from the medium and is labeled the "work horse". Any time you see a mechanical device such as a turbine or compressor adding or removing energy from a medium you are looking at an application of a constant specific entropy process was involved. If the change in specific entropy (s) is constant then the change in specific enthalpy heat (hq) and specific internal energy heat (uq) are constant. Therefore the heat added to or removed from this process is zero (Q = 0). .

To begin the discussion lets look at the constant entropy data of steam/water. Remember that hq = uq = 491.69 R x s.

Temp °R	Pressure psia	Volume ft³/lbm	Entropy Btu/lbm-°R	Enthalpy Btu/lbm	uq / hq Btu/lbm	hw Btu/lbm	Pv work Btu/lbm	Tmh °R
585.7	2	136.820	1.5500	899.4	762.1	137.2	50.6	580.23
671.6	14.7	22.947	1.5500	1011.5	762.1	249.4	62.4	652.58
787.5	100	4.221	1.5500	1145.6	762.1	383.5	78.1	739.10
1162.1	800	0.786	1.5500	1340.8	762.1	578.7	116.3	865.03
1494.0	2400	0.334	1.5500	1485.5	762.1	723.4	148.1	958.39

Observation from the table

1. Entropy is constant = 1.55 (Btu / lbm-R)
2. Hq / uq is constant = 762.1 (Btu/lbm) = 491.69 x 1.55 = 762..1 (Btu/lbm)

Example below is a constant specific entropy process with energy removed (-) from the medium.

H2O	Temp	Pressure	Volume	Entropy	Enthalpy	uq / hq	hw	Pv work	Tmh
491.69	°R	psia	ft³/lbm	Btu/lbm-°R	Btu/lbm	Btu/lbm	Btu/lbm	Btu/lbm	°R
1- Initial	1494.0	2400	0.334	1.5500	1485.5	762.1	723.4	148.1	958.39
2- Final	585.7	2	136.820	1.5500	899.4	762.1	137.2	50.6	580.23
(2 - 1)	-908.3	-2398.0	136.486	0.0000	-586.2	0.0	-586.2	-97.5	-378.16
1	lbm /hr		136.49	0.0000	-586.2	0.0	-586.2	-97.5	

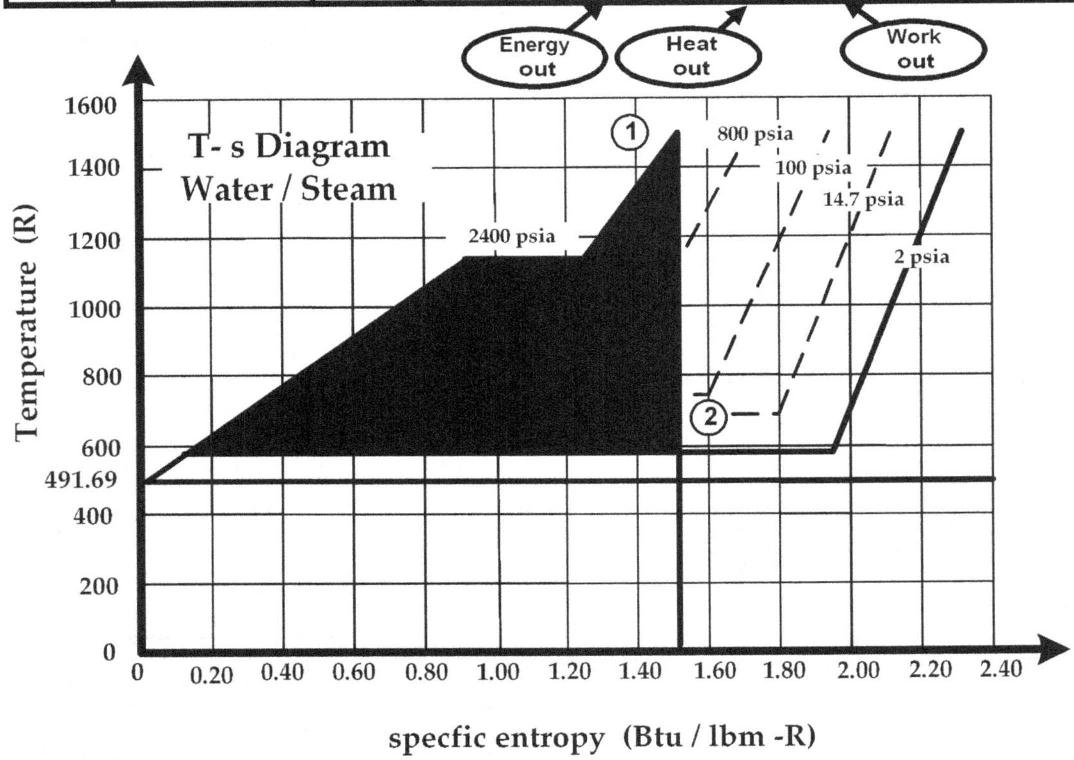

Observations

1. Specific entropy is constant @ 1.5500 Btu/ lbm- R
Specific enthalpy heat = specific internal energy heat @ 762.1 Btu / lbm

2. Energy removed from media = (black area) x (mass)
black area = change in specific enthalpy (h)

3. Work removed from media = (black area above 491.69 R) x (mass)
black area above 491.69 R = change in specific enthalpy work(hw)

4. No Heat removed from media = (black area below 491.69 R) x (mass) = 0
black area below 491.69 R = change in specific enthapy heat (hq) =0

5. Change in specific internal energy (not shown)= h - Pv work = - 488.7 Btu/lbm

Observations: Results of (2 – 1)

H2O	Temp	Pressure	Volume	Entropy	Enthalpy	uq / hq	hw	Pv work	Tmh
491.69	°R	psia	ft³/lbm	Btu/lbm-°R	Btu/lbm	Btu/lbm	Btu/lbm	Btu/lbm	°R
1- Initial	671.6	14.7	22.947	1.5500	1011.5	762.1	249.4	62.4	652.58
2- Final	1162.1	800	0.786	1.5500	1340.8	762.1	578.7	116.3	865.03
(2 - 1)	490.5	785.3	-22.161	0.0000	329.3	0.0	329.3	53.9	212.45
1	lbm /hr		-22.16	0.0000	329.3	0.0	329.3	53.9	

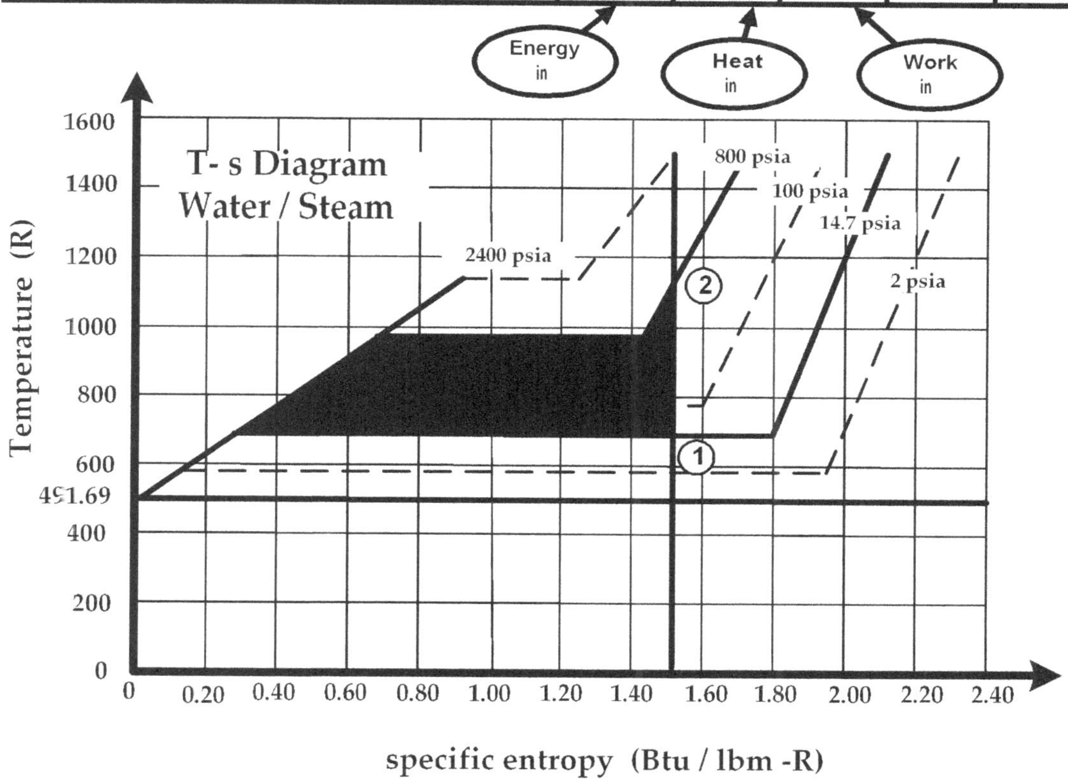

Observations

1. Specific entropy is constant @ 1.5500 Btu/ lbm- R
Specific enthalpy heat = specific internal energy heat @ 762.1 Btu / lbm

2. Energy added to media = (black area) x (mass)
black area = change in specific enthalpy (h)

3. Work added to media = (black area above 491.69 R) x (mass)
black area above 491.69 R = change in specific enthalpy work (hw)

4. No Heat added to media = (black area below 491.69 R) x (mass) = 0
black area below 491.69 R = change in specific enthalpy heat (hq) = 0

5. Change in specific internal energy (not shown)= h - Pv work = 275.4 Btu / lbm

Using constant pressure and constant volume processes to achieve the same initial and final results.

H2O	Temp	Pressure	Volume	Entropy	Enthalpy	uq / hq	hw	Pv work	Tmh
491.69	°R	psia	ft³/lbm	Btu/lbm-°R	Btu/lbm	Btu/lbm	Btu/lbm	Btu/lbm	°R
1- Initial	671.6	14.7	22.947	1.5500	1011.5	762.1	249.4	62.4	652.58
2- Final	671.6	14.7	0.786	0.3539	208.2	174.0	34.2	2.1	588.25
(2 - 1)	0.0	0.0	-22.161	-1.1961	-803.3	-588.1	-215.2	-60.3	-64.34
1	lbm /hr		-22.16	-1.1961	-803.3	-588.1	-215.2	-60.3	

H2O	Temp	Pressure	Volume	Entropy	Enthalpy	uq / hq	hw	Pv work	Tmh
491.69	°R	psia	ft³/lbm	Btu/lbm-°R	Btu/lbm	Btu/lbm	Btu/lbm	Btu/lbm	°R
1- Initial	671.6	14.7	0.786	0.3539	208.2	174.0	34.2	2.1	588.25
2- Final	1162.4	800	0.786	1.5502	1341.0	762.2	578.8	116.3	865.05
(2 - 1)	490.8	785.3	0.000	1.1963	1132.8	588.2	544.6	114.2	276.80
1	lbm /hr		0.00	1.1963	1132.8	588.2	544.6	114.2	

Combined results

1	lbm /hr		-22.161	0.0000	329.3	0.0	329.3	53.9	

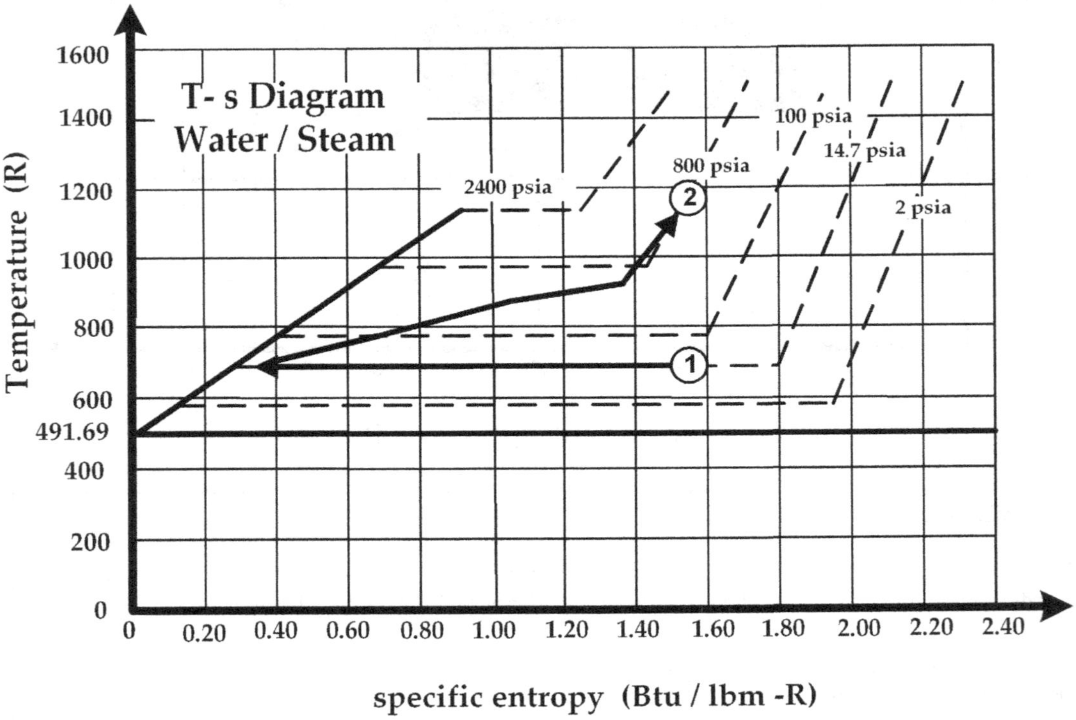

3.5 Constant Specific Enthalpy... h1 = h2

The constant enthalpy process is considered the "throttling" process and one of the most confusing processes in thermodynamics. At first glance the constant enthalpy process looks like no energy is released from the medium because the enthalpy change is zero. As the medium passes through an orifice the pressure decreases but the enthalpy stays the same. It was like nothing happened to the energy level of the medium. It looks like the perfect "environmental friendly" process. This process destroys more available energy to do work than any process in industry. The constant enthalpy process takes work energy from the medium and converts in to heat energy. The overall energy seems unchanged but this process destroys the work energy added to the medium such as a steam turbine cycle or a refrigeration system. Another name for the constant enthalpy process is "zero percent efficiency nozzle". The "orifice effect" may not seem obvious but it is a key component in the inefficiency in other processes. Below is data at a constant enthalpy.

Temp	Pressure	Volume	Entropy	Enthalpy	uq / hq	hw	Pv work	Tmh
°R	psia	ft^3/lbm	Btu/lbm-°R	Btu/lbm	Btu/lbm	Btu/lbm	Btu/lbm	°R
1076.2	2	320.460	2.2034	1345.0	1083.4	261.6	118.6	610.42
1077.9	14.7	43.590	1.9836	1345.0	975.3	369.7	118.5	678.06
1089	100	6.399	1.7734	1345.0	872.0	473.0	118.4	758.43
1168.8	800	0.792	1.5536	1345.0	763.9	581.1	117.2	865.73
1300	2400	0.261	1.4490	1345.0	712.5	632.5	115.9	928.23

Observations:
The specific enthalpy is constant @ 1345.0 Btu / lbm
The specific enthalpy work increases with pressure.
The specific enthalpy heat decreases with pressure.

Example below is a constant specific enthalpy process with no energy removed from or added to the medium. The work energy is converted to heat energy.

H2O	Temp	Pressure	Volume	Entropy	Enthalpy	uq / hq	hw	Pv work	Tmh
491.69	°R	psia	ft³/lbm	Btu/lbm-°R	Btu/lbm	Btu/lbm	Btu/lbm	Btu/lbm	°R
1- Initial	1168.8	800	0.792	1.5536	1345.0	763.9	581.1	117.2	865.73
2- Final	1077.9	14.7	43.590	1.9836	1345.0	975.3	369.7	118.5	678.06
(2 - 1)	-90.9	-785.3	42.80	0.4300	0.0	211.4	-211.4	1.3	-187.67
1	lbm /hr		42.798	0.4300	0.0	211.4	-211.4	1.3	

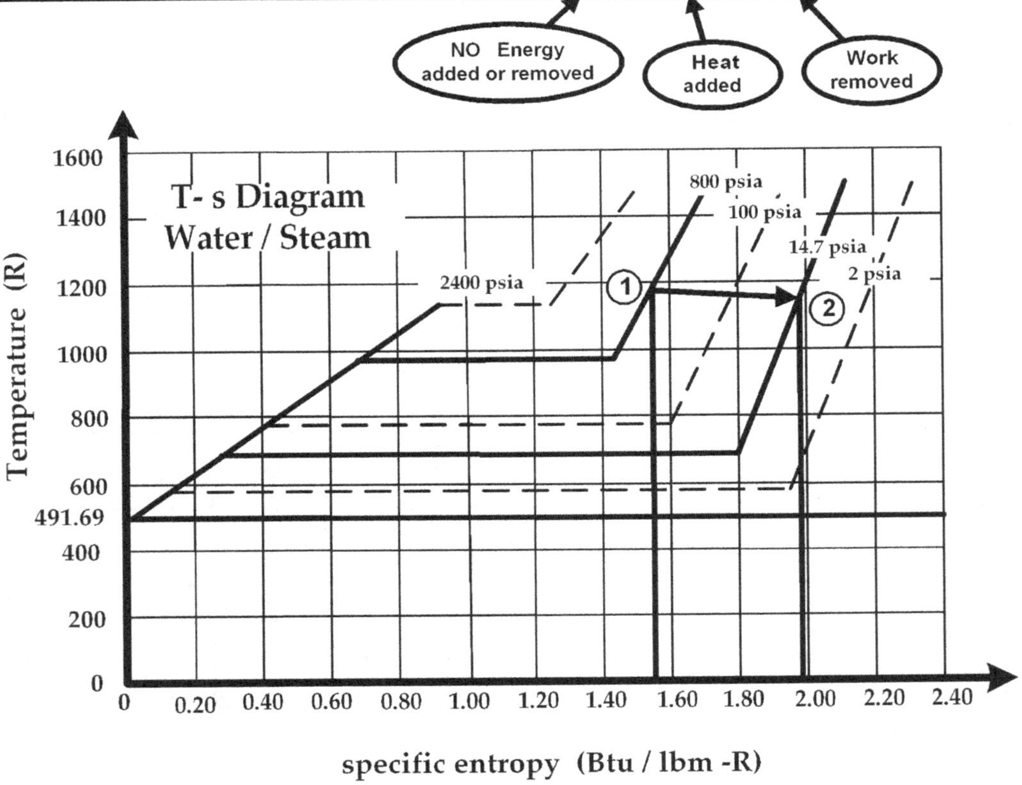

Observations

1. Specific enthalpy is constant @ 1345.0 Btu/ lbm
2. Energy added to media = zero
3. Work removed from media = change in specific enthalpy work (hw) = -211.4 Btu/ lbm
4. Heat added to media = change in specific enthalpy heat (hq) = + 211.4 Btu/ lbm
5. The work was converted to heat

Using constant specific entropy and constant pressure processes to achieve the same initial and final results:

H_2O	Temp	Pressure	Volume	Entropy	Enthalpy	uq / hq	hw	Pv work	Tmh
491.69	°R	psia	ft³/lbm	Btu/lbm-°R	Btu/lbm	Btu/lbm	Btu/lbm	Btu/lbm	°R
1- Initial	1168.6	800	0.792	1.5535	1344.9	763.8	581.1	117.2	865.72
2- Final	671.64	14.7	23.011	1.5535	1013.9	763.8	250.1	62.6	652.66
(2 - 1)	-497.0	-785.3	22.22	0.0000	-331.0	0.0	-331.0	-54.6	-213.07
1	lbm /hr		22.219	0.0000	-331.0	0.0	-331.0	-54.6	

H_2O	Temp	Pressure	Volume	Entropy	Enthalpy	uq / hq	hw	Pv work	Tmh
491.69	°R	psia	ft³/lbm	Btu/lbm-°R	Btu/lbm	Btu/lbm	Btu/lbm	Btu/lbm	°R
1- Initial	671.64	14.7	23.011	1.5535	1013.9	763.8	250.1	62.6	652.66
2- Final	1077.9	14.7	43.590	1.9836	1345.0	975.3	369.7	118.5	678.06
(2 - 1)	406.3	0.0	20.58	0.4301	331.1	211.5	119.6	56.0	25.40
1	lbm /hr		20.579	0.4301	331.10	211.5	119.6	56.0	

Combined Results

1	lbm /hr		42.798	0.4300	0.00	211.4	-211.4	1.3	

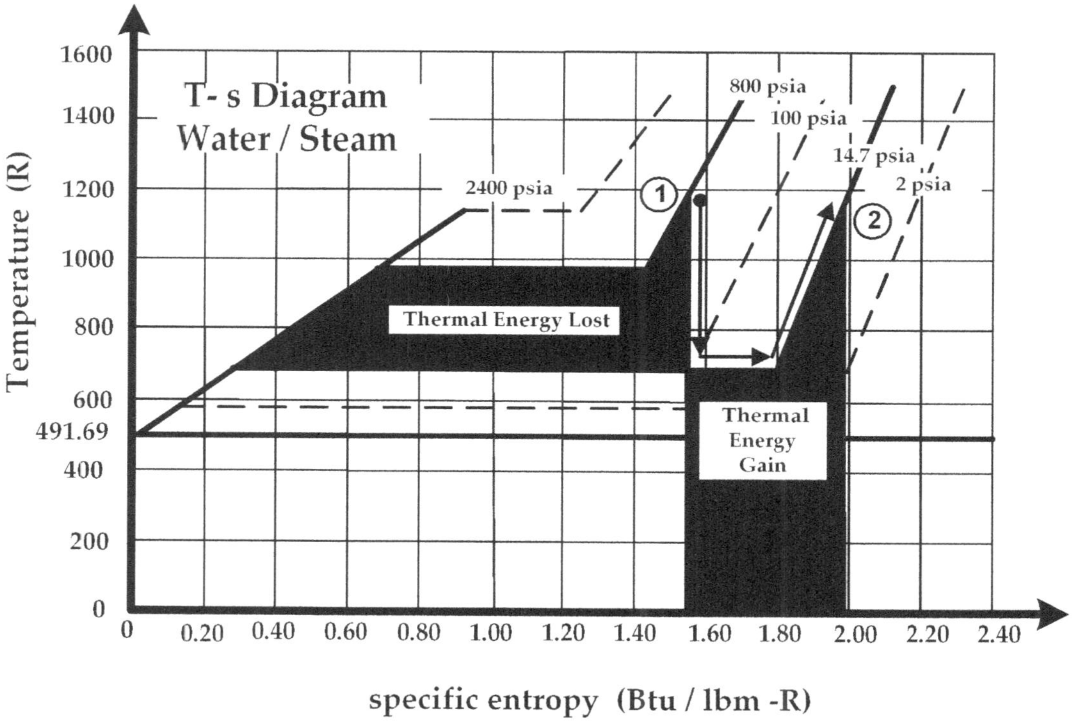

3.6 Constant Temperature... T1 = T2

The constant temperature process is the most overrated and confusing process in thermodynamics. This process is sometimes called the isothermal process and is used in most books for the understanding of thermodynamics. A closer look at the constant temperature process shows a work energy consuming process similar to the constant enthalpy process. This process converts work energy to heat energy but also requires more energy to maintain a constant temperature. Below is a table of constant temperature data:

Temp	Pressure	Volume	Entropy	Enthalpy	uq / hq	hw	Pv work	Tmh
°R	psia	ft³/lbm	Btu/lbm-°R	Btu/lbm	Btu/lbm	Btu/lbm	Btu/lbm	°R
1300	2	387.140	2.2962	1455.0	1129.0	326.0	143.2	633.66
1300	14.7	52.627	2.0759	1454.5	1020.7	433.8	143.1	700.66
1300	100	7.691	1.8626	1451.2	915.8	535.4	142.3	779.13
1300	800	0.913	1.6169	1423.0	795.0	628.0	135.1	880.08
1300	2400	0.261	1.4490	1345.0	712.5	632.5	115.9	928.23

Observations:
The constant temperature data:

1. The temperature is constant
2. Increase the pressure and the specific enthalpy decreases
3. Increase the specific enthalpy work and the specific enthalpy heat decreases

Let's take a closer look at the data between two constant temperature points.

H2O	Temp	Pressure	Volume	Entropy	Enthalpy	uq / hq	hw	Pv work	Tmh
491.69	°R	psia	ft³/lbm	Btu/lbm-°R	Btu/lbm	Btu/lbm	Btu/lbm	Btu/lbm	°R
1- Initial	1300	800	0.913	1.6169	1423.0	795.0	628.0	135.1	880.08
2- Final	1300	14.7	52.627	2.0759	1454.5	1020.7	433.8	143.1	700.66
(2 - 1)	0.0	-785.3	51.71	0.4590	31.5	225.7	-194.2	8.1	-179.42
1	lbm /hr		51.714	0.4590	31.50	225.69	-194.19	8.05	

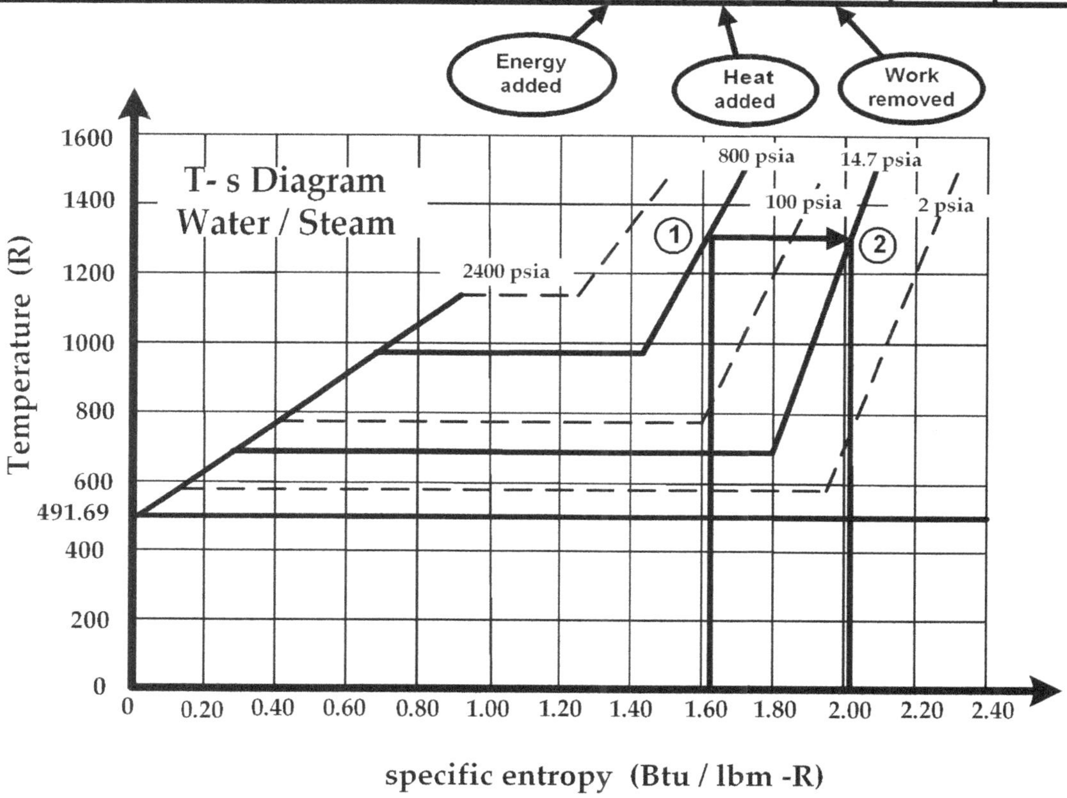

Observations

1. Temperature is constant @ 1300 R
2. Energy added to media = + 31.5 Btu/ lbm
3. Work removed from media = change in specific enthalpy work (hw) = -194.19 Btu/ lbm
4. Heat added to media = change in specific enthalpy heat (hq) = + 225.69 Btu/ lbm
5. The work was converted to heat and requires + 31.5 Btu/ lbm from an outside source.

Using constant specific entropy and constant pressure processes to achieve the same initial and final results

H2O	Temp	Pressure	Volume	Entropy	Enthalpy	uq / hq	hw	Pv work	Tmh
491.69	°R	psia	ft³/lbm	Btu/lbm-°R	Btu/lbm	Btu/lbm	Btu/lbm	Btu/lbm	°R
1- Initial	1300	800	0.913	1.6169	1423.0	795.0	628.0	135.1	880.08
2- Final	671.64	14.7	24.186	1.6169	1056.5	795.0	261.5	65.8	653.41
(2 - 1)	-628.4	-785.3	23.27	0.0000	-366.5	0.0	-366.5	-69.3	-226.67
1	lbm /hr		23.273	0.0000	-366.50	0.00	-366.50	-69.29	

H2O	Temp	Pressure	Volume	Entropy	Enthalpy	uq / hq	hw	Pv work	Tmh
491.69	°R	psia	ft³/lbm	Btu/lbm-°R	Btu/lbm	Btu/lbm	Btu/lbm	Btu/lbm	°R
1- Initial	671.64	14.7	24.186	1.6169	1056.5	795.0	261.5	65.8	653.41
2- Final	1300	14.7	52.627	2.0759	1454.5	1020.7	433.8	143.1	700.66
(2 - 1)	628.4	0.0	28.44	0.4590	398.0	225.7	172.3	77.3	47.25
1	lbm /hr		28.441	0.4590	398.00	225.69	172.31	77.35	
				Results	31.50	225.69	-194.19		

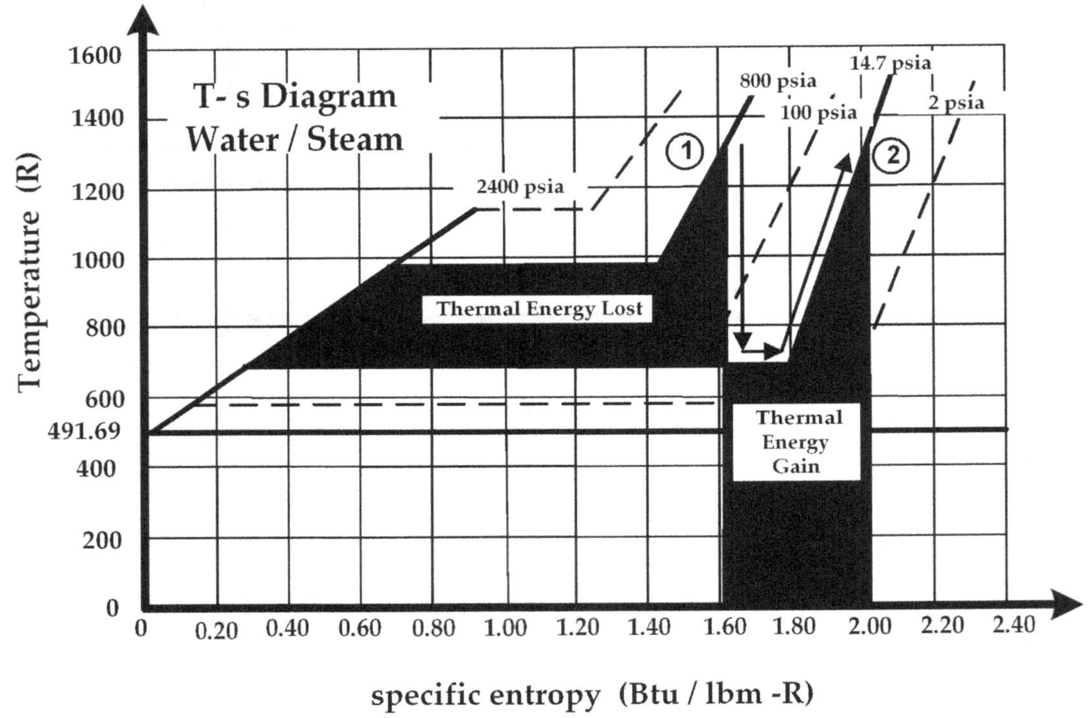

Using constant specific enthalpy and constant pressure processes to achieve the same initial and final results:

H2O	Temp	Pressure	Volume	Entropy	Enthalpy	uq / hq	hw	Pv work	Tmh
491.69	°R	psia	ft³/lbm	Btu/lbm-°R	Btu/lbm	Btu/lbm	Btu/lbm	Btu/lbm	°R
1- Initial	1300	800	0.913	1.6169	1423.0	795.0	628.0	135.1	880.08
2- Final	1237	14.7	50.065	2.0511	1423.0	1008.5	414.5	136.2	693.77
(2 - 1)	-63.0	-785.3	49.15	0.4342	0.0	213.5	-213.5	1.1	-186.31
1	lbm /hr		49.152	0.4342	0.00	213.49	-213.49	1.08	

H2O	Temp	Pressure	Volume	Entropy	Enthalpy	uq / hq	hw	Pv work	Tmh
491.69	°R	psia	ft³/lbm	Btu/lbm-°R	Btu/lbm	Btu/lbm	Btu/lbm	Btu/lbm	°R
1- Initial	1237	14.7	50.065	2.0511	1423.0	1008.5	414.5	136.2	693.77
2- Final	1300	14.7	52.627	2.0759	1454.5	1020.7	433.8	143.1	700.66
(2 - 1)	63.0	0.0	2.56	0.0248	31.5	12.2	19.3	7.0	6.89
1	lbm /hr		2.562	0.0248	31.50	12.19	19.31	6.97	

Notice how the constant enthalpy process destroys available work and then it is required to add energy from a constant pressure process. Maybe if the process was reversed it would be useful.

H2O	Temp	Pressure	Volume	Entropy	Enthalpy	uq / hq	hw	Pv work	Tmh
491.69	°R	psia	ft³/lbm	Btu/lbm-°R	Btu/lbm	Btu/lbm	Btu/lbm	Btu/lbm	°R
1- Initial	1300	14.7	52.627	2.0759	1454.5	1020.7	433.8	143.1	700.66
2- Final	1300	800	0.913	1.6169	1423.0	795.0	628.0	135.1	880.08
(2 - 1)	0.0	785.3	-51.71	-0.4590	-31.5	-225.7	194.2	-8.1	179.42
1	lbm /hr		-51.714	-0.4590	-31.50	-225.69	194.19	-8.05	

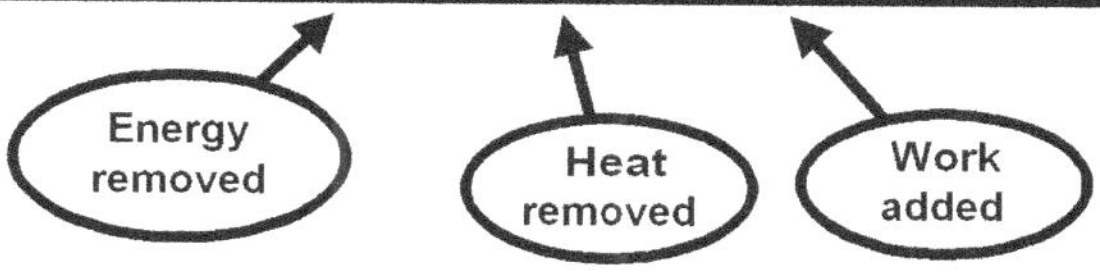

Energy and heat was removed from the process and it created work. Remember one of the statements from the beginning of the book...

"Work can create heat... but heat can not create Work."

3.7 Constant Internal Energy… u1 = u2

H2O	Temp	Pressure	Volume	Entropy	Enthalpy	uq / hq	hw	Pv work	Tmh	Int Egy	uw
491.69	°R	psia	ft³/lbm	Btu/lbm-°R	Btu/lbm	Btu/lbm	Btu/lbm	Btu/lbm	°R	Btu/lbm	Btu/lbm
1- Initial	1231	2400.0	0.228	1.3986	1281.3	687.7	593.6	101.2	916.13	1180.1	492.4
2- Final	1075	800.0	0.696	1.4983	1283.0	736.7	546.3	102.9	856.30	1180.1	443.4
(2 - 1)	-155	-1600.0	0.468	0.0997	1.7	49.0	-47.3	1.7	-59.83	0.0	-49.1
1	lbm /hr		0.468	0.0997	1.70	49.02	-47.32	1.73		-0.03	-49.06

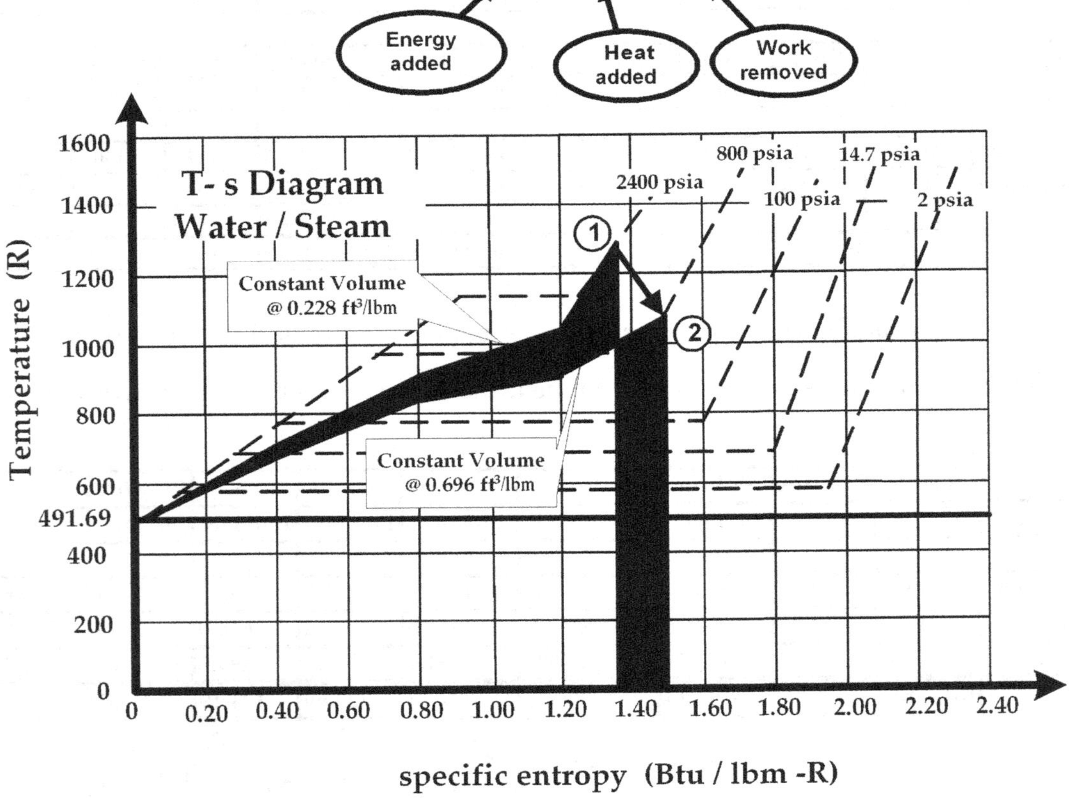

Observations

1. Specific internal energy constant @ 1180.1 Btu / lbm
2. Change in uw = -49.06 change in uq = 49.02 Btu / lbm
3. Change in h = change in Pv work = 1.70 Btu / lbm
4. The black area between below the constant volume line @ 0.228 and above the constant specific volume line @ 0.696 ft^3 / lbm is specific internal enegy removed
5. The black area to the right between below the constant volume line @ 0.696 ft ^{3} / lbm is the specific internal energy added
6. Constant internal energy is simular to the constant temperature process, not useful.

Using constant specific entropy and constant specific volume processes to achieve the same initial and final results.

H2O 491.69	Temp °R	Pressure psia	Volume ft³/lbm	Entropy Btu/lbm-°R	Enthalpy Btu/lbm	uq / hq Btu/lbm	hw Btu/lbm	Pv work Btu/lbm	Tmh °R	Int Egy Btu/lbm	uw Btu/lbm
1- Initial	1231	2400.0	0.228	1.3987	1281.4	687.7	593.7	101.2	916.14	1180.2	492.4
2- Final	950	625.5	0.696	1.3987	1162.0	687.7	474.3	80.5	830.77	1081.5	393.7
(2 - 1)	-280	-1774.5	0.468	0.0000	-119.4	0.0	-119.4	-20.7	-85.36	-98.7	-98.7
1	lbm /hr		0.468	0.0000	-119.40	0.00	-119.40	-20.69		-98.71	-98.71

H2O 491.69	Temp °R	Pressure psia	Volume ft³/lbm	Entropy Btu/lbm-°R	Enthalpy Btu/lbm	uq / hq Btu/lbm	hw Btu/lbm	Pv work Btu/lbm	Tmh °R	Int Egy Btu/lbm	uw Btu/lbm
1- Initial	950	625.5	0.696	1.3987	1162.0	687.7	474.3	80.5	830.77	1081.5	393.7
2- Final	1076	800.0	0.696	1.4985	1283.3	736.8	546.5	103.0	856.39	1180.3	443.5
(2 - 1)	125	174.5	0.000	0.0998	121.3	49.1	72.2	22.5	25.62	98.8	49.8
1	lbm /hr		0.000	0.0998	121.30	49.07	72.23	22.46		98.84	49.76

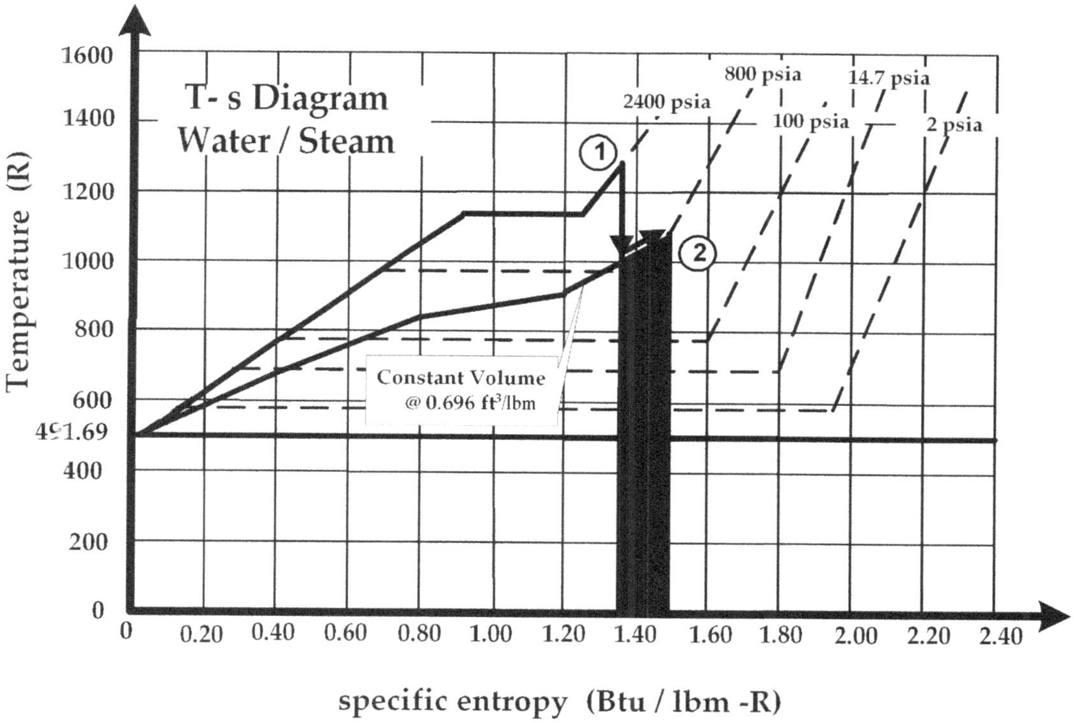

Below is data at a constant specific internal energy at 1180.1 Btu/lbm

Temp °R	Pressure psia	Volume ft³/lbm	Entropy Btu/lbm-°R	Enthalpy Btu/lbm	uq / hq Btu/lbm	hw Btu/lbm	Pv work Btu/lbm	Tmh °R	Int Egy Btu/lbm	uw Btu/lbm
950	2	282.850	2.1440	1284.8	1054.2	230.6	104.7	599.25	1180.1	126.0
952	14.7	38.461	1.9241	1284.7	946.1	338.6	104.6	667.69	1180.1	234.0
968	100	5.642	1.7145	1284.5	843.0	441.5	104.4	749.20	1180.1	337.1
1075	800	0.696	1.4983	1283.0	736.7	546.3	102.9	856.30	1180.1	443.4
1231	2400	0.228	1.3986	1281.3	687.7	593.6	101.2	916.13	1180.1	492.4

3.8 Constant Pv work... (0.185) P1v1 = (0.185) P2v2

H2O	Temp	Pressure	Volume	Entropy	Enthalpy	uq / hq	hw	Pv work	Tmh	uw
491.69	°R	psia	ft³/lbm	tu/lbm-°I	Btu/lbm	Btu/lbm	Btu/lbm	Btu/lbm	°R	Btu/lbm
1- Initia	1493.4	2400	0.333	1.5497	1485.1	762.0	723.1	148.0	958.31	575.1
2- Final	1344.2	14.7	54.422	2.0928	1476.7	1029.0	447.7	148.0	705.61	299.7
(2 - 1)	-149.2	-2385.3	54.089	0.5431	-8.4	267.0	-275.4	0.0	-252.70	-275.4
1	lbm /hr		54.089	0.5431	-8.40	267.04	-275.44	0.00		-275.4

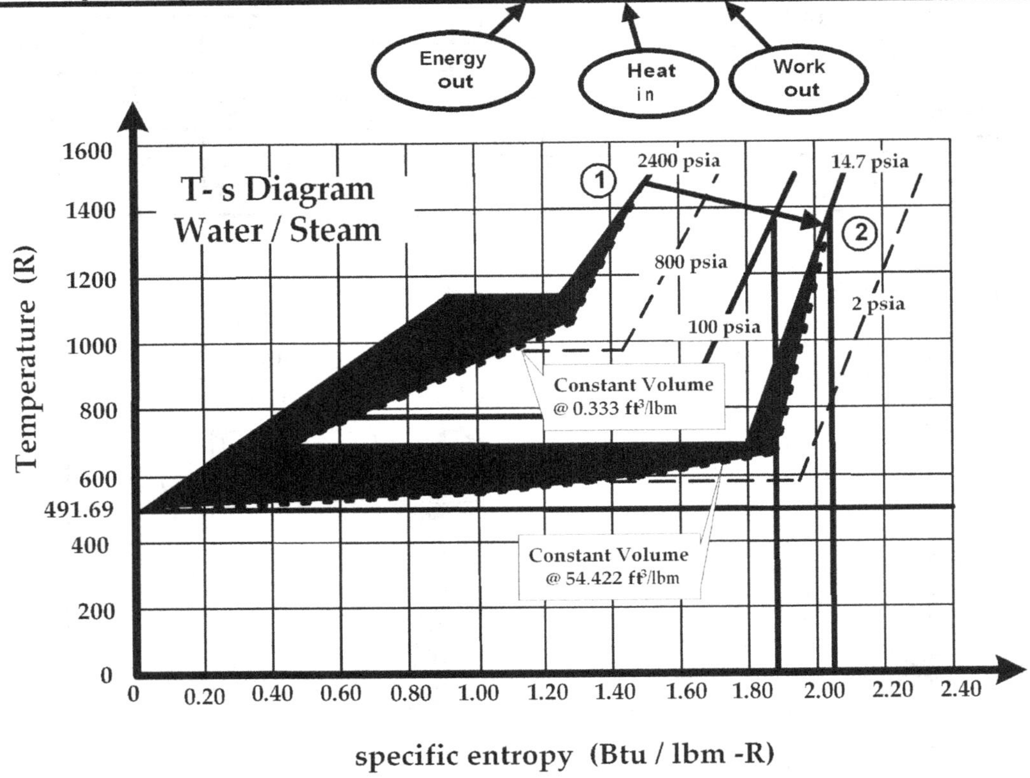

Observations

1. Specific Pv work constant @ 148 Btu / lbm
2. Change in uw = change in hw = -275.44 Btu / lbm
3. Change in h = change in u = - 8.4 Btu / lbm
4. Initial Pv work is the black area between below the constant pressure line @ 2400 and above the constant specific volume line @ 0.333 ft^3 / lbm
5. Final Pv work is the black area between below the constant pressure line @ 14.7 and above the constant specific volume line @ 54.422 ft^3 / lbm
6. The initial and final black area are equal (Pv work)1 = (Pv work)2

Using constant specific entropy and constant specific enthalpy processes to achieve the same initial and final results

H2O	Temp	Pressure	Volume	Entropy	Enthalpy	uq / hq	hw	Pv work	Tmh	uw
491.69	°R	psia	ft³/lbm	tu/lbm-°I	Btu/lbm	Btu/lbm	Btu/lbm	Btu/lbm	°R	Btu/lbm
1- Initial	1493.4	2400	0.333	1.5497	1485.1	762.0	723.1	148.0	958.31	575.1
2- Final	1474.6	2269.3	0.348	1.5497	1476.8	762.0	714.8	146.2	952.96	568.6
(2 - 1)	-18.8	-130.7	0.015	0.0000	-8.3	0.0	-8.3	-1.8	-5.36	-6.5
1	lbm /hr		0.015	0.0000	-8.30	0.00	-8.30	-1.82		-6.5

H2O	Temp	Pressure	Volume	Entropy	Enthalpy	uq / hq	hw	Pv work	Tmh	uw
491.69	°R	psia	ft³/lbm	tu/lbm-°I	Btu/lbm	Btu/lbm	Btu/lbm	Btu/lbm	°R	Btu/lbm
1- Initial	1474.6	2269.3	0.348	1.5497	1476.8	762.0	714.8	146.2	952.96	568.6
2- Final	1344.2	14.7	54.422	2.0928	1476.7	1029.0	447.7	148.0	705.61	299.7
(2 - 1)	-130.4	-2254.6	54.074	0.5431	-0.1	267.0	-267.1	1.8	-247.35	-269.0
1	lbm /hr		54.074	0.5431	-0.10	267.04	-267.14	1.82		-269.0

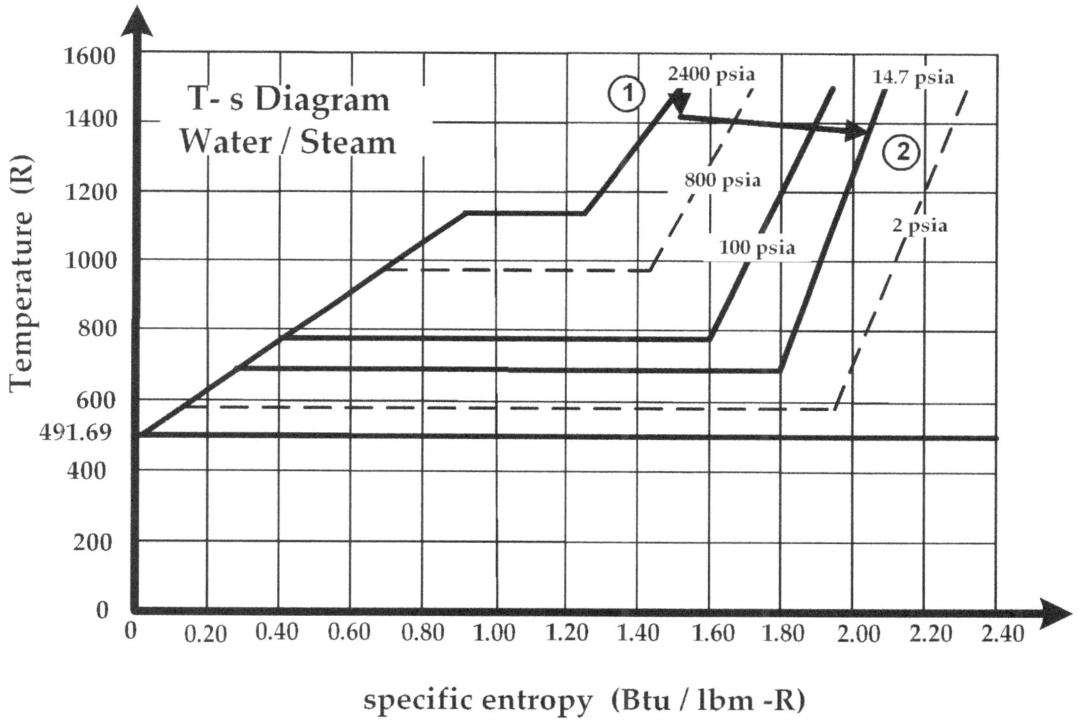

Below is data at a constant specific Pv work at 148.0 Btu/lbm

Temp	Pressure	Volume	Entropy	Enthalpy	uq / hq	hw	Pv work	Tmh	uw
°R	psia	ft³/lbm	tu/lbm-°I	Btu/lbm	Btu/lbm	Btu/lbm	Btu/lbm	°R	Btu/lbm
1343.1	2	400.000	2.3126	1476.7	1137.1	339.6	148.0	638.55	191.6
1344.2	14.7	54.422	2.0928	1476.7	1029.0	447.7	148.0	705.61	299.7
1350.9	100	8.000	1.8822	1477.2	925.5	551.7	148.0	784.83	403.7
1401.7	800	1.000	1.6596	1480.7	816.0	664.7	148.0	892.20	516.7
1493.4	2400	0.333	1.5497	1485.1	762.0	723.1	148.0	958.31	575.1

3.9 Constant Enthalpy Work ... hw1 = hw2

H2O	Temp	Pressure	Volume	Entropy	Enthalpy	uq / hq	hw	pv	Tmh	uw
491.69	°R	psia	ft³/lbm	Btu/lbm-°R	Btu/lbm	Btu/lbm	Btu/lbm	Btu/lbm	°R	Btu/lbm
1- Initial	1187.7	14.7	48.061	2.0310	1398.6	998.6	400.0	130.7	688.63	269.3
2- Final	1121.8	2400	0.093	1.1017	941.7	541.7	400.0	41.5	854.76	358.5
(2 - 1)	-65.9	2385.3	-47.97	-0.9293	-456.9	-456.9	0.0	-89.2	166.13	89.3
1	lbm /hr		-47.968	-0.9293	-456.9	-456.9	0.0	-89.2		89.3

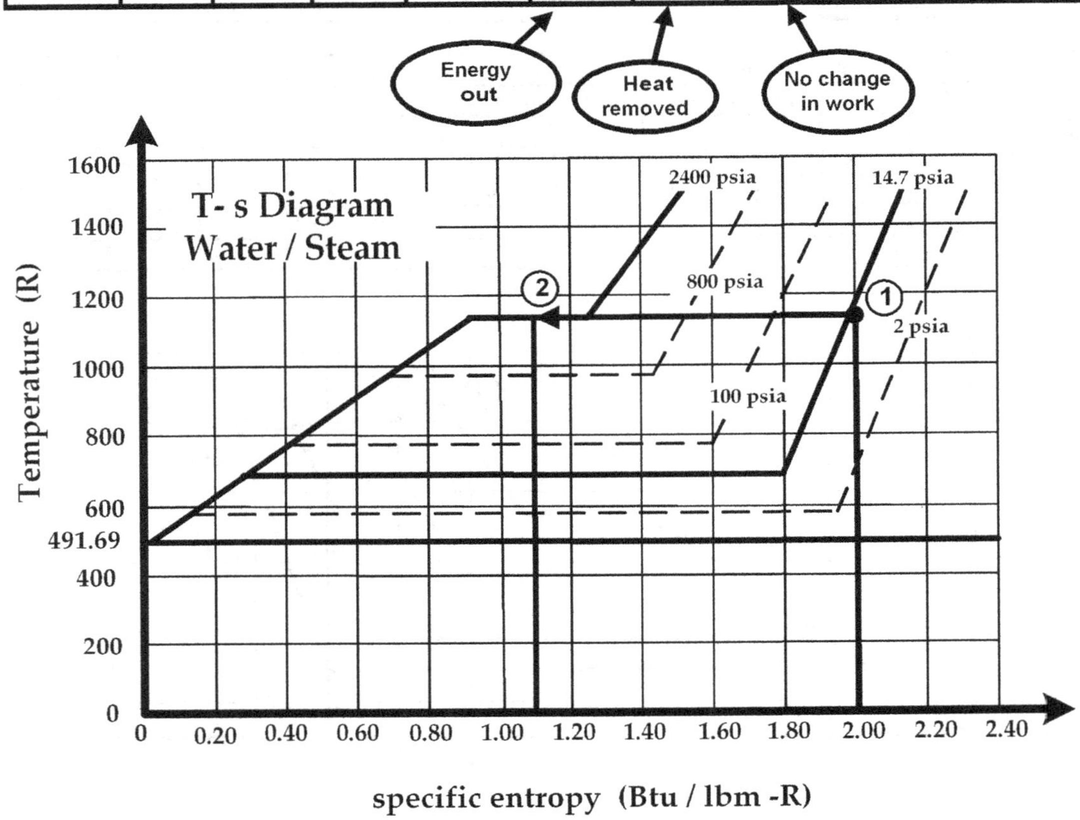

Observations

1. Specific enthalpy work constant @ 400 Btu / lbm
2. Change in uw = + 89.3 change in Pv work = - 89.3 Btu / lbm
3. Change in h = change in hq work = - 456.9 Btu / lbm
4. To locate each constant hw point
   a. Change h_2 to the desired (941.7)
   b. Calculate change in h = 941.7 - 1398.6 = - 456.9
   c. Divide - 456.9 / 491.69 = - 0.92924 (change in specific entropy)
   d. final entropy = 2.0310 - 0.92924 = 1.10175
5. Constant enthapy work process can be the perfect heat sink, removing only heat. Just remember that some other process has to absorb this heat.

Using constant pressure and constant specific entropy processes to achieve the same initial and final results

H2O	Temp	Pressure	Volume	Entropy	Enthalpy	uq / hq	hw	pv	Tmh	uw
491.69	°R	psia	ft³/lbm	Btu/lbm-°R	Btu/lbm	Btu/lbm	Btu/lbm	Btu/lbm	°R	Btu/lbm
(1) 1- Initial	1187.7	14.7	48.061	2.0310	1398.6	998.6	400.0	130.7	688.63	269.3
2- Final	671.64	14.7	14.641	1.1017	710.4	541.7	168.7	39.8	644.85	128.9
(2 - 1)	-516.1	0.0	-33.42	-0.9293	-688.2	-456.9	-231.2	-90.9	-43.78	-140.4
1	lbm /hr		-33.420	-0.9293	-688.2	-456.9	-231.2	-90.9		-140.4

H2O	Temp	Pressure	Volume	Entropy	Enthalpy	uq / hq	hw	pv	Tmh	uw
491.69	°R	psia	ft³/lbm	Btu/lbm-°R	Btu/lbm	Btu/lbm	Btu/lbm	Btu/lbm	°R	Btu/lbm
1- Initial	671.64	14.7	14.641	1.1017	710.4	541.7	168.7	39.8	644.85	128.9
(2) 2- Final	1121.8	2400	0.093	1.1017	941.7	541.7	400.0	41.5	854.76	358.5
(2 - 1)	450.2	2385.3	-14.55	0.0000	231.3	0.0	231.3	1.6	209.91	229.6
1	lbm /hr		-14.548	0.0000	231.3	0.0	231.3	1.6		229.6

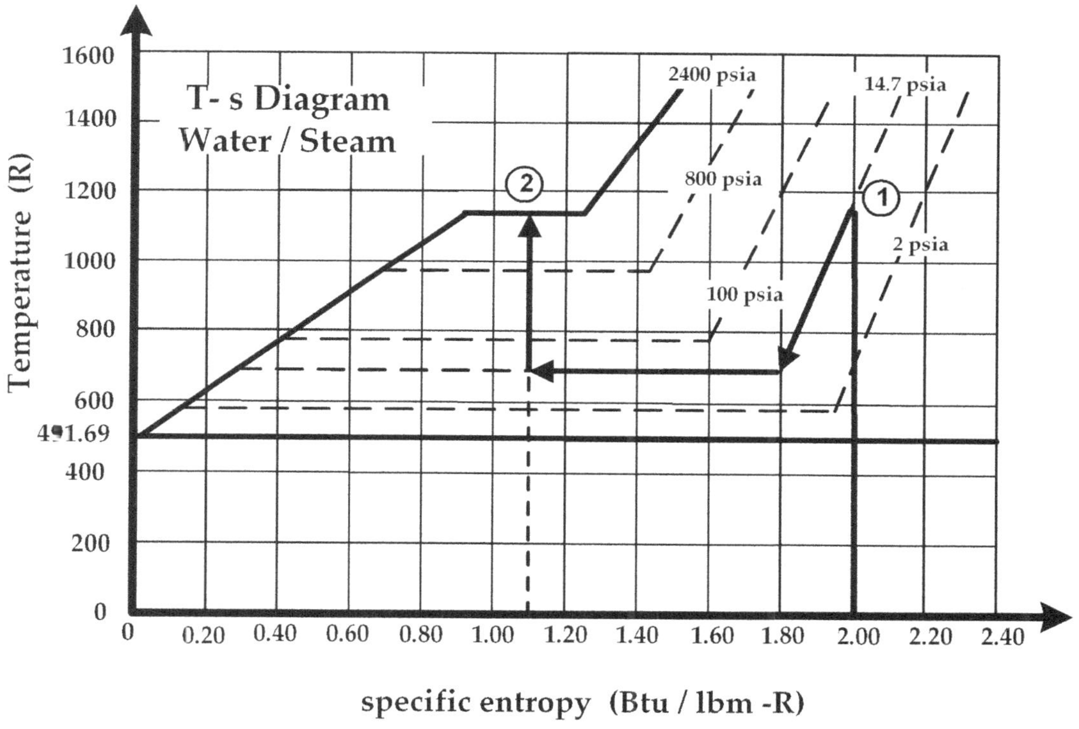

Below is an example of constant enthalpy work process adding energy with no change in work.

H2O	Temp	Pressure	Volume	Entropy	Enthalpy	uq / hq	hw	pv	Tmh	uw
491.69	°R	psia	ft³/lbm	Btu/lbm-°R	Btu/lbm	Btu/lbm	Btu/lbm	Btu/lbm	°R	Btu/lbm
1- Initial	789.7	100	4.449	1.6060	1189.7	789.7	400.0	82.3	740.78	317.7
2- Final	1187.7	14.7	48.061	2.0310	1398.6	998.6	400.0	130.7	688.63	269.3
(2 - 1)	398.0	-85.3	43.61	0.4250	208.9	209.0	-0.1	48.4	-52.16	-48.5
1	lbm /hr		43.612	0.4250	208.9	209.0	-0.1	48.4		-48.5

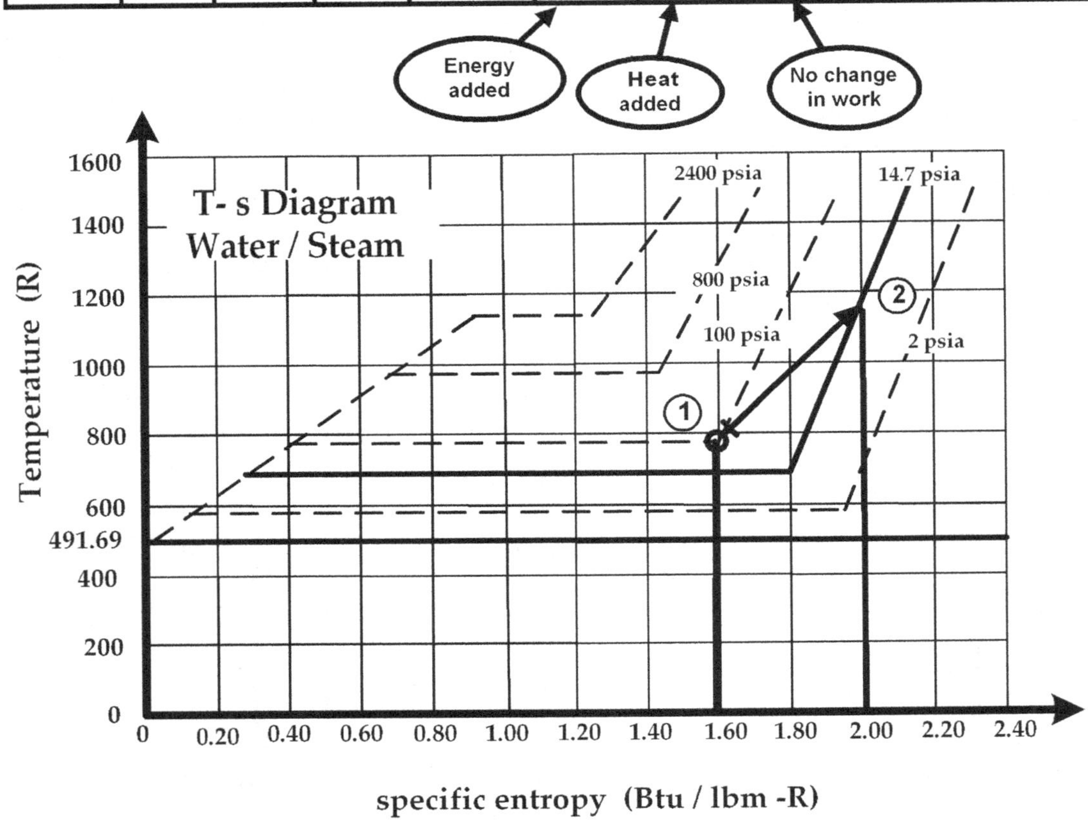

Below is a table with enthalpy work constant at 400 Btu / lbm.

Temp	Pressure	Volume	Entropy	Enthalpy	uq / hq	hw	pv	Tmh	uw
°R	psia	ft³/lbm	Btu/lbm-°R	Btu/lbm	Btu/lbm	Btu/lbm	Btu/lbm	°R	Btu/lbm
1376.6	5	163.960	2.2240	1493.5	1093.5	400.0	151.7	671.54	248.3
1187.7	14.7	48.061	2.0310	1398.6	998.6	400.0	130.7	688.63	269.3
789.7	100	4.449	1.6060	1189.7	789.7	400.0	82.3	740.78	317.7
977.94	800	0.404	1.2047	992.3	592.3	400.0	59.8	823.72	340.2
1121.8	2400	0.093	1.1017	941.7	541.7	400.0	41.5	854.76	358.5

3.10 Cycle... the multi- processes

The "cycle" is a series of interconnecting processes of one medium that returns to its original state. The cycle does not define the source that supplied the energy to or the sink that absorbed the energy from each process that is defined by the "system". One example of a water steam cycle is the Rankine cycle which is made up of two constant pressure and two constant entropy processes. Below is an illustration of a Rankine cycle.

Rankine Cycle

Pump

H2O	Temp	Pressure	Volume	Entropy	Enthalpy	uq / hq	hw	Pv work	Tmh
491.69	°R	psia	ft³/lbm	Btu/lbm-°R	Btu/lbm	Btu/lbm	Btu/lbm	Btu/lbm	°R
1- Initial	585.7	2	0.016	0.1751	94.09	86.10	7.99	0.01	537.29
2- Final	1121.8	2400	0.028	0.9033	719.15	444.15	275.00	12.38	796.12
(2 - 1)	536.1	2398.0	0.01	0.7282	625.07	358.05	267.01	12.38	258.8
1	lbm /hr		0.012	0.7282	625.07	358.05	267.01	12.38	

Boiler

H2O	Temp	Pressure	Volume	Entropy	Enthalpy	uq / hq	hw	Pv work	Tmh
491.69	°R	psia	ft³/lbm	Btu/lbm-°R	Btu/lbm	Btu/lbm	Btu/lbm	Btu/lbm	°R
1- Initial	1121.8	2400	0.028	0.9033	719.15	444.15	275.00	12.38	796.12
2- Final	2716.2	2400	0.674	1.9207	2243.60	944.39	1299.21	299.31	1168.12
(2 - 1)	1594.4	0.0	0.65	1.0174	1524.45	500.24	1024.21	286.93	372.0
1	lbm /hr		0.646	1.0174	1524.45	500.24	1024.21	286.93	

Turbine

H2O	Temp	Pressure	Volume	Entropy	Enthalpy	uq / hq	hw	Pv work	Tmh
491.69	°R	psia	ft³/lbm	Btu/lbm-°R	Btu/lbm	Btu/lbm	Btu/lbm	Btu/lbm	°R
1- Initial	2716.2	2400	0.674	1.9207	2243.60	944.39	1299.21	299.31	1168.12
2- Final	585.7	2	173.710	1.9207	1116.50	944.39	172.11	64.27	581.30
(2 - 1)	-2130.5	-2398.0	173.04	0.0000	-1127.10	0.00	-1127.10	-235.04	-586.8
1	lbm /hr		173.036	0.0000	-1127.10	0.00	-1127.10	-235.04	

Condenser

H2O	Temp	Pressure	Volume	Entropy	Enthalpy	uq / hq	hw	Pv work	Tmh
491.69	°R	psia	ft³/lbm	Btu/lbm-°R	Btu/lbm	Btu/lbm	Btu/lbm	Btu/lbm	°R
1- Initial	585.7	2	173.710	1.9207	1116.50	944.39	172.11	64.27	581.30
2- Final	585.7	2	0.016	0.1751	94.09	86.10	7.99	0.01	537.29
(2 - 1)	0.0	0.0	-173.69	-1.7456	-1022.42	-858.29	-164.13	-64.27	-44.0
1	lbm /hr		-173.694	-1.7456	-1022.42	-858.29	-164.13	-64.27	

Rankine cycle
Basic Power plant cycle
Operating between 2 and 2400 psia

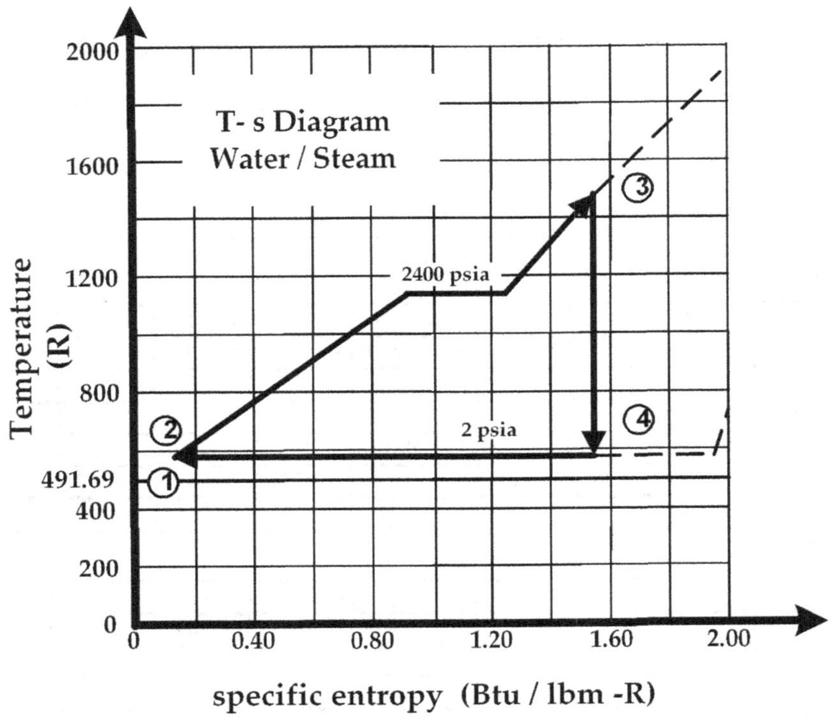

	H2O	Temp	Pressure	Volume	Entropy	Enthalpy	uq / hq	hw	Pv work	Tmh
	491.69	°R	psia	ft³/lbm	Btu/lbm-°R	Btu/lbm	Btu/lbm	Btu/lbm	Btu/lbm	°R
Point "1"	Liquid	585.7	2	0.016	0.1751	94.09	86.09	8.00	0.01	537.35
Point "2"	Liquid	586.8	2400	0.016	0.1751	101.26	86.09	15.17	7.16	578.30
Point "3"	Superheat	1460	2400	0.322	1.5346	1462.80	754.55	708.25	142.96	953.21
Point "4"	Vapor	585.7	2	135.290	1.5346	890.33	754.55	135.78	50.06	580.17

Below is an example of a reverse Rankine cycle using water / steam. This cycle is used in air conditioning but typically not with water as a medium.

Compressor

H2O	Temp	Pressure	Volume	Entropy	Enthalpy	uq / hq	hw	Pv work	Tmh
491.69	°R	psia	ft³/lbm	tu/lbm-°R	Btu/lbm	Btu/lbm	Btu/lbm	Btu/lbm	°R
1- Initial	500	0.12331	2300.200	2.0585	1029.20	1012.14	17.06	52.47	499.98
2- Final	1255.5	14.7	50.819	2.0585	1432.20	1012.14	420.06	138.20	695.75
(2 - 1)	755.5	14.6	-2249.38	0.0000	403.00	0.00	403.00	85.73	195.8
1	lbm		-2249.381	0.0000	403.00	0.00	403.00	85.73	

Condenser

H2O	Temp	Pressure	Volume	Entropy	Enthalpy	uq / hq	hw	Pv work	Tmh
491.69	°R	psia	ft³/lbm	tu/lbm-°R	Btu/lbm	Btu/lbm	Btu/lbm	Btu/lbm	°R
1- Initial	1255.5	14.7	50.819	2.0585	1432.20	1012.14	420.06	138.20	695.75
2- Final	671.64	14.7	0.017	0.3124	180.30	153.59	26.71	0.05	577.18
(2 - 1)	-583.9	0.0	-50.80	-1.7461	-1251.90	-858.55	-393.35	-138.16	-118.6
1	lbm		-50.802	-1.7461	-1251.90	-858.55	-393.35	-138.16	

Orifice throttling

H2O	Temp	Pressure	Volume	Entropy	Enthalpy	uq / hq	hw	Pv work	Tmh
491.69	°R	psia	ft³/lbm	tu/lbm-°R	Btu/lbm	Btu/lbm	Btu/lbm	Btu/lbm	°R
1- Initial	671.64	14.7	0.017	0.3124	180.30	153.59	26.71	0.05	577.18
2- Final	500	0.12331	387.420	0.3607	180.30	177.37	2.93	8.84	499.81
(2 - 1)	-171.6	-14.6	387.40	0.0484	0.00	23.78	-23.78	8.79	-77.4
1	lbm		387.403	0.0484	0.00	23.78	-23.78	8.79	

Evaporator

H2O	Temp	Pressure	Volume	Entropy	Enthalpy	uq / hq	hw	Pv work	Tmh
491.69	°R	psia	ft³/lbm	tu/lbm-°R	Btu/lbm	Btu/lbm	Btu/lbm	Btu/lbm	°R
1- Initial	500	0.12331	387.420	0.3607	180.30	177.37	2.93	8.84	499.81
2- Final	500	0.12331	2300.200	2.0585	1029.20	1012.14	17.06	52.47	499.98
(2 - 1)	0.0	0.0	1912.78	1.6978	848.90	834.77	14.13	43.64	0.2
1	lbm		1912.780	1.6978	848.90	834.77	14.13	43.64	

A cycle doesn't show the work energy and heat energy transferred to each process.

4 Weber's Laws of Thermodynamics

"If it's not work, it must be heat"

Fred J. Weber

4.1 Introduction to Weber's Thermodynamics Laws

The laws of thermodynamics were developed to explain the transfer of energy between two mediums. The Weber definition of thermodynamics is "the science that deals with work and heat energy within a medium and the interaction of heat and work between two mediums". The term used to describe this transfer of energy is called a "System".

The process defines what energy, work and heat was added or removed. The system defines where the energy, work and heat came from or went to. The laws of thermodynamics were written for a "system" not for a process or cycle.

The system defines the energy balance. Below is a system diagram showing the energy, work and heat being transferred between two mediums.

Wrench Time System Diagram

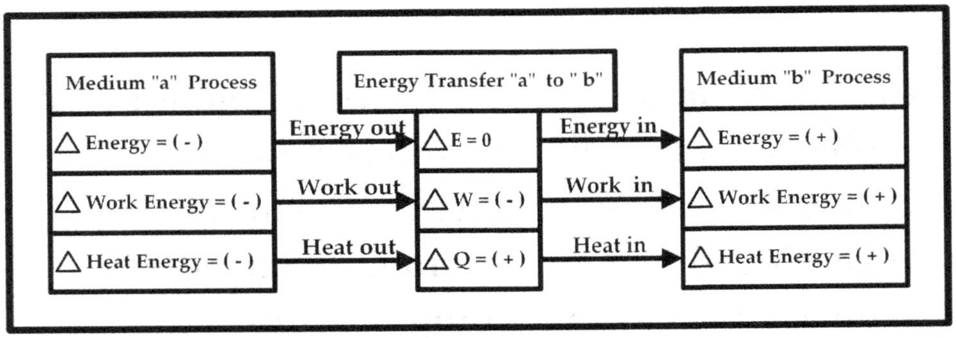

To understand these Weber's Laws of Thermodynamics we need to first review the following information:

Step 1

Recognize that all energy has two components and they are "work energy" and "heat energy".

Step 2

Grasp the definition of these words:

Energy (E)… capacity available and unavailable for performing work.

Work (W)… change in work energy ; change in energy available for performing work

Heat (Q)… change in heat energy ; change in energy unavailable for performing work

Work Energy (Ew) … capacity of energy available for performing work

Heat energy (Eq) … capacity of energy unavailable for performing work

Step 3

Establish a set of equations used to define these laws:

Weber's General Energy Equations

E = Ew + Eq : all energy has two components work energy and heat energy

W = Δ Ew : work is equal to the change in work energy

Q = Δ Eq : heat is equal to the change in heat energy

Δ E = Δ Ew + Δ Eq = W + Q : change in energy is equal to work plus heat

Step 4

Understand the common sense sign convention for transferring energy between two mediums:

(+) = added or in....energy in, energy added, work in, work added, heat in, heat added

(-) = removed or out...energy out, energy removed, work out, work removed, heat removed

(1) = initial condition

(2) = final condition

Using a common sense sign convention:

(+ Δ E) = Ein (the change in energy is positive (+) then energy is added to the medium)

(- Δ E) = Eout (the change in energy is negative (-) then energy is removed from the medium)

(+) W = Win = (+ Δ Ew) (the change in work energy is positive (+) then work is added to the medium (Win)

(-) W = Wout = **(- Δ Ew)** (the change in work energy is negative (-) then work is removed from the medium (Wout).

Using this same sign convention for heat:

(+) Q = Qin = (+ Δ Eq) (the change in heat energy is positive (+) then heat is added to the medium (Qin)

(-) Q = Qout = (- Δ Eq) (the change in heat energy is negative (-) then heat is removed from the medium (Qout).

Steps 5

Display the information. Develop a system that shows the path and data of the flow of energy, work and heat. Remember the example of the isolated hot and cold container. By themselves they are two separate mediums with nothing to change their original state. At the moment the hot container comes in contact with the cold container the energy transfer begins. The energy from the hot transfers to the cold container and is displayed below using a "Weber System Diagram":

H2O	Temp	Pressure	Volume	Entropy	Enthalpy	uq / hq	hw	Pv work	Tmh
491.69	°R	psia	ft³/lbm	Btu/lbm-°R	Btu/lbm	Btu/lbm	Btu/lbm	Btu/lbm	°R
1- Initial	580	14.7	0.016	0.1653	88.43	81.29	7.13	0.04	534.84
2- Final	560	14.7	0.016	0.1303	68.44	64.05	4.39	0.04	525.40
(2 - 1)	-20.0	0.0	0.00	-0.0351	-19.98	-17.24	-2.74	0.00	-9.4
2	lbm		0.000	-0.0701	-39.96	-34.48	-5.48	0.00	

Energy	Heat	Work
0.00	1.24	-1.25

H2O	Temp	Pressure	Volume	Entropy	Enthalpy	uq / hq	hw	Pv work	Tmh
491.69	°R	psia	ft³/lbm	Btu/lbm-°R	Btu/lbm	Btu/lbm	Btu/lbm	Btu/lbm	°R
1- Initial	540	14.7	0.016	0.0939	48.47	46.19	2.27	0.04	515.89
2- Final	560	14.7	0.016	0.1303	68.44	64.05	4.39	0.04	525.40
(2 - 1)	20.0	0.0	0.00	0.0363	19.98	17.86	2.12	0.00	9.5
2	lbm		0.000	0.0727	39.96	35.72	4.24	0.00	

4.2 Weber's 1st Law of Thermodynamics

Weber's 1st law of thermodynamics is identical to the traditional first law of thermodynamics that states "the energy can not be created or destroyed". The Weber's interpretation of this law refers to the transfer of energy between two mediums and not on the change of energy within a process or cycle. This law defines the energy interaction between two mediums which is displayed at the top part of the system diagram.

Weber's 1st Law of Thermodynamics

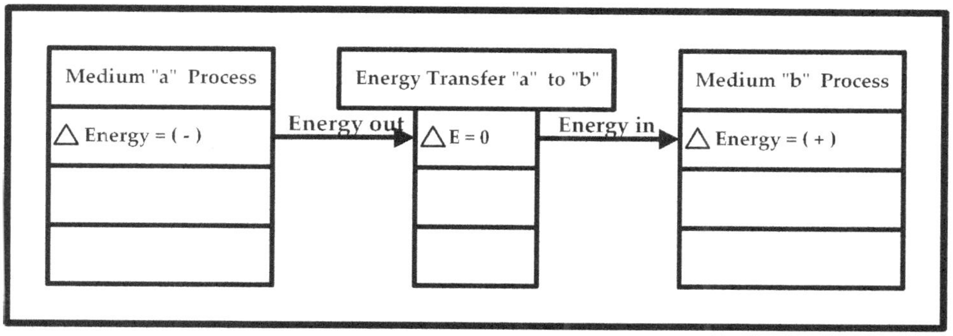

This law states the change in energy medium "a" plus the change in energy medium "b" must equal to zero. For this equation to be true one of the energy changes must be (-) and the other one needs to be (+).

Assume medium "a" is the source (removing energy) then the change in energy will be (-). This sets medium "b" to be the sink and the change in energy will be (+). Therefore the energy removed from medium "a" plus energy added to medium "b" is equal to zero. In the discussion of energy we are talking about all energy which includes thermal, kinetic, potential and etc. The focus in this section is just the change in thermal energy as a result the change in energy = change in thermal energy = change in enthalpy.

Below are the equations that support the 1st law

Weber's 1st law of Thermodynamics

$\Delta Ea + \Delta Eb = 0$

Let the : Δ Energy = ΔThermal Energy = ΔEnthalpy

$\Delta E = \Delta TE = \Delta H$

The energy equation goes from

$\Delta Ea + \Delta Eb = 0$ **TO** $\Delta Ha + \Delta Hb = 0$

$(Ha_2 - Ha_1) + (Hb_2 - Hb_1) = 0$

Note: (2) final reading and (1) initial reading

Substituting to specfic enthalpy

$ma(ha_2 - ha_1) + mb(hb_2 - hb_1) = 0$

For this equation to equal zero either

$ha_2 - ha_1$ must be (-) and $hb_2 - hb_1$ must be (+) **OR**

$ha_2 - ha_1$ must be (+) and $hb_2 - hb_1$ must be (-)

Selecting medium "a" as the source (user's choice) then

$ha_2 - ha_1$ must be (-) and $hb_2 - hb_1$ must be (+)

Summary : $\Delta Ha + \Delta Hb = 0$

Medium "a"		**Medium "b"**
$\Delta Ha = ma(ha_2 - ha_1) = (-)$	⇨	$\Delta Hb = mb(hb_2 - hb_1) = (+)$
Energy is removed from medium "a"	⇨	Energy is added to medium "b"
Energy is going out of medium "a"	⇨	Energy is going into medium "b"

Maintain the selection of medium "a" as the source of energy to medium "b" thoughout the discussion of the laws of thermodynamics

4.3 Weber's 2nd Law Thermodynamics

The Weber's 2nd Law of Thermodynamics is based on two statements:

"Work can create heat... but heat can not create Work."

"The interaction of two mediums "can not" increase the available work of the system"

These statements are based on the assumption that...

" all energy has two components and they are "work energy" and "heat energy "

These statements can be represented on the bottom portion of the System Diagram below:

Weber's 2nd Law of Thermodynamics

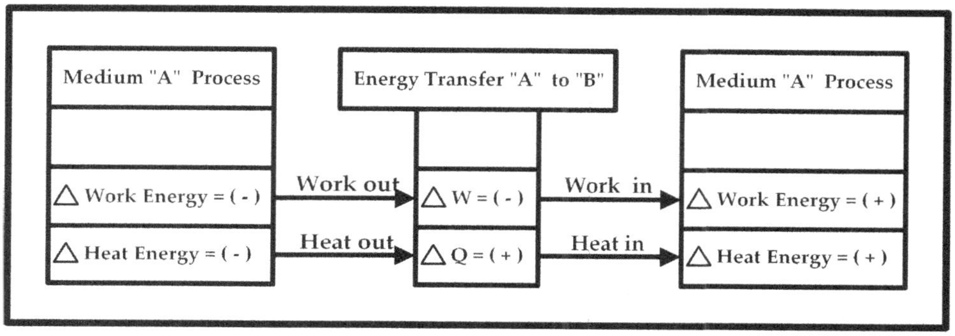

Assume medium "a" is the source and the change in energy is (-). Then medium "b" is the sink and the change in energy will be (+). The energy available to do work removed from medium "a" plus energy available to do work added to medium "b" must be (-). No work can be created.

Therefore the net work is equal to or less than zero.

Total change in work = (-) = (Wa + Wb)

As a result the net heat is equal to or greater than zero

Total change in heat = (+) = (Qa + Qb)

Because the net work plus net heat is equal to zero.

Total change in energy = (0) = (Wa + Wb) + (Qa + Qb)

Below are the equations that support the Weber's 2nd law of thermodynamics:

Weber's second law of Thermodynamics is based on:

" all energy has two components: work energy and heat energy "

Energy = Work Energy + Heat Energy

$E = Ew + Eq$

Ew = work energy

Eq = heat energy

The change in energy is equal to the change in work energy plus the change in heat energy:

$\Delta E = \Delta Ew + \Delta Eq = (Ew_2 - Ew_1) + (Eq_2 - Eq_1)$

$\Delta Ew = (Ew_2 - Ew_1)$ the change in work energy

$\Delta Eq = (Eq_2 - Eq_1)$ the change in heat energy

$\Delta Ew = m(ew_2 - ew_1)$ mass times the change in specific work energy

$\Delta Eq = m(eq_2 - eq_1)$ mass times the change in specific heat energy

Note: (2) final reading and (1) initial reading

Work and Heat

W = work added to or removed from the medium

$W = \Delta Ew$ (work is equal to the change in work energy)

Q = heat added to or removed from the medium

$Q = \Delta Eq$ (heat is equal to the change in heat energy)

The Energy equation becomes:

$$\Delta E = \Delta Ew + \Delta Eq = W + Q$$

Construct energy equations for medium "a" and "b" :

Medium "a"

$\Delta Ea = \Delta Ewa + \Delta Eqa = Wa + Qa$

$\Delta Ewa = ma(ewa_2 - ewa_1)$

$\Delta Eqa = ma(eqa_2 - eqa_1)$

$\Delta Ea = ma[(ewa_2 - ewa_1) + (eqa_2 - eqa_1)]$

$\Delta Ea = Wa + Qa$

$Wa = ma(ewa_2 - ewa_1)$

$Qa = ma(eqa_2 - eqa_1)$

$Ea = Wa + Qa$

$$\Delta Ea = \Delta Ewa + \Delta Eqa = Wa + Qa$$

Medium "b"

$\Delta Eb = \Delta Ewb + \Delta Eqb = Wb + Qb$

$\Delta Ewb = mb(ewb_2 - ewb_1)$

$\Delta Eqb = mb(eqb_2 - eqb_1)$

$\Delta Eb = mb[(ewb_2 - ewb_1) + (eqb_2 - eqb_1)]$

$\Delta Eb = Wb + Qb$

$Wb = mb(ewb_2 - ewb_1)$

$Qb = mb(eqb_2 - eqb_1)$

$Eb = Wb + Qb$

$$\Delta Eb = \Delta Ewb + \Delta Eqb = Wb + Qb$$

Applying the 1st law energy equation

page 2

$\triangle$Ea + $\triangle$Eb = 0

$\triangle$Ewa + $\triangle$Eqa+ $\triangle$Ewb + $\triangle$Eqa = 0

$ma\,[\,(ewa_2 - ewa_1) + (eqa_2 - eqa_1)\,] + mb\,[\,(ewb_2 - ewb_1) + (eqb_2 - eqb_1)\,] = 0$

Wa + Qa + Wb + Qb = 0

Weber's second law statements are:

"Work can create heat... but heat can not create Work."

"The interaction of two medium "can not" increase the available work of the system"

The equation that supports these statements... **Wa + Wb ≤ 0**

Work from medium "a" plus the work from medium "b" must be less than or equal to zero.

Wa + Qa + Wb + Qb = 0

If Wa + Wb must be (-) or equal to zero

then Qa + Qb must be (+) or equal to zero

and Wa + Wb = - (Qa + Qb)

Examples

If Wa + Wb = 0 then Qa + Qb =0

If Wa + Wb = - 5 then Qa + Qb = +5

If Wa + Wb = - 32 then Qa + Qb = +32

Substituting to specfic energy

If $ma\,(ewa_2 - ewa_1) + mb\,[\,(ewb_2 - ewb_1) = -32.74$

then $ma\,(eqa_2 - eqa_1) + mb\,(eqb_2 - eqb_1) = +32.74$

Review

Wa + Wb = ma (ewa2 - ewa1) + mb [(ewb2 - ewb1) = (-) or 0

Qa + Qb = ma (eqa2 - eqa1) + mb (eqb2 - eqb1) = (+) or 0

Wa + Wb = - (Qa + Qb)

Wa + Qa + Wb + Qb = 0

page 3

Assume all the energy transferd between medium "a" and b" is thermal energy only:

E = Te = H = (m)(h) (Energy = Thermal Energy = Enthalpy = mass x specific enthalpy)

Construct thermal energy equations for medium "a" and "b" :

Medium "a"

$\Delta Ha = (Ha_2 - Ha_1)$

$\Delta Ha = \Delta Hwa + \Delta Hqa = Wa + Qa$

$\Delta Ha = ma (ha_2 - ha_1)$

$\Delta Ha = \Delta Hwa + \Delta Hqa$

$\Delta Hwa = ma (hwa_2 - hwa_1)$

$\Delta Hqa = ma (hqa_2 - hqa_1)$

$\Delta Ha = ma [(hwa_2 - hwa_1) + (hqa_2 - hqa_1)]$

$Wa = ma (hwa_2 - hwa_1)$

$Qa = ma (hqa_2 - hqa_1)$

$\Delta Ha = Wa + Qa$

Medium "b"

$\Delta Hb = (Hb_2 - Hb_1)$

$\Delta Hb = \Delta Hwb + \Delta Hqb = Wb + Qb$

$\Delta Hb = mb (hb_2 - hb_1)$

$\Delta Hb = \Delta Hwb + \Delta Hqb$

$\Delta Hwb = mb (hwb_2 - hwb_1)$

$\Delta Hqb = mb (hqb_2 - hqb_1)$

$\Delta Hb = mb [(hwb_2 - hwb_1) + (hqb_2 - hqb_1)]$

$Wb = mb (hwb_2 - hwb_1)$

$Qb = mb (hqb_2 - hqb_1)$

$\Delta Hb = Wb + Qb$

Applying the equation from the first law

$\Delta Ha + \Delta Hb = 0$

$\Delta Hwa + \Delta Hqa + \Delta Hwb + \Delta Hqb = 0$

$ma [(hwa_2 - hwa_1) + (hqa_2 - hqa_1)] + mb [(hwb_2 - hwb_1) + (hqb_2 - hqb_1)] = 0$

$Wa + Qa + Wb + Qb = 0$

"The interaction of two mediums "can not" increase the available work of the system"

If $Wa + Wb$ can not be (+)

then $Wa + Wb \leq 0$ (must be (-) or equal to zero)

then $(Hwa_2 - Hwa_1) + (Hwb_2 - Hwb_1) \leq 0$ (must be (-) or equal to zero)

then $ma(hwa_2 - hwa_1) + mb(hwb_2 - hwb_1) \leq 0$ (must be (-) or equal to zero)

Review: $Wa + Wb \leq 0$ and $Qa + Qb \geq 0$ with $Wa + Qa + Wb + Qb = 0$

If $Wa + Wb = 0$ then $Qa + Qb = 0$

If $Wa + Wb = - 32.74$ then $Qa + Qb = +32.74$

If $(Hwa_2 - Hwa_1) + (Hwb_2 - Hwb_1) = - 32.74$

then $(Hqa_2 - Hqa_1) + (Hqb_2 - Hqb_1) = + 32.74$

If $ma (hwa_2 - hwa_1) + mb [(hwb_2 - hwb_1) = - 32.74$

then $ma (hqa_2 - hqa_1) + mb [(hqb_2 - hqb_1) = + 32.74$

Summary

page 4

"Work can create heat... but heat can not create Work."

"Work can create heat... "

Wa + Wb can be (-) and create Qa + Qb to be (+)

"... but heat can not create Work."

Qa + Qb **can not** be (-) andcreate Wa + Wbto be (+)

Where:

Wa + Wb = net work of the system

Qa + Qb= net heat of the system

Two statements are used to describe the second law of thermodynamics:

1. "All energy is converted to Heat"

   Wa + Wb = (-) or zero

   Qa + Qb i= (+) or zero

net work is decreasing while net heat of a system is always increasing

2. "The entropy of a system is always increasing"

   Qa + Qb is to be (+) or zero

   $Qa + Qb = (Hqa_2 - Hqa_1) + (Hqb_2 - Hqb_1) = (+)$ or zero

   Hq = Tr x S

   Hq = Enthalpy Heat

   Tr = Reference temperature (491.69 °R)

   S = Entropy

   $(Hqa_2 - Hqa_1) + (Hqb_2 - Hqb_1) = (+)$ or zero

   $(Tr) (Sa_2 - Sa_1) + (Sb_2 - Sb_1) = (+)$ or zero

Divide by sides by (Tr)

$(Sa_2 - Sa_1) + (Sb_2 - Sb_1) = [(+)$ or zero $] / Tr$

The change in entropy of a system $(Sa_2 - Sa_1) + (Sb_2 - Sb_1)$ is increasing

For specific entropy

$ma (hqa_2 - hqa_1) + mb (hqb_2 - hqb_1) = (+)$ or zero

$(Tr) [ma (sa_2 - sa_1) + mb (sb_2 - sb_1)]= (+)$ or zero

Divide by sides by (Tr)

The change in specific entropy of a system $ma (sa_2 - sa_1) + mb (sb_2 - sb_1)$ is increasing

Weber's second law of Thermodynamics meets these two statement requirements

4.4 Weber's 3rd Law Thermodynamics

Weber's Third law of Thermodynamics

Requirements for a self acting thermal interaction between two mediums

Assume

- medium "a" is the source and medium "b" is the sink.
- all the energy transferred between mediums "a" and b" is thermal energy only:
- E = Te = H = (m)(h) (Energy = Thermal Energy = Enthalpy = mass x specific enthalpy)
- requirements below will be maintain throughout the interaction medium "a" & medium "b"

Prerequisite for a thermal self acting system between medium "a" as the source and medium "b" as the sink.

1. $ha_1 > hb_1$
specific enthalpy of the source("a") is greater than the specific enthalpy of the sink ("b")

2. $hwa_1 > hwb_1$
specific enthalpy work of the source("a") is greater than
the specific enthalpy work of the sink ("b")

3. $hqa_1 > hqb_1$
specific enthalpy heat of the source("a") is greater than
the specific enthalpy heat of the sink ("b")

4. $Tmha_1 > Tmhb_1$
mean enthalpy temperature of the source("a") is greater than
the mean enthalpy temperature of the sink ("b")

1. $ha_1 > hb_1$ 3rd law... page 2

specific enthalpy of the source("a") is greater than the specific enthalpy of the sink ("b")

$$\Delta Ha + \Delta Hb = 0$$

$$\Delta Ha = ma\,(ha_2 - ha_1) = (-)$$

$$\Delta Hb = mb\,(hb_2 - hb_1) = (+)$$

$$ma\,(ha_2 - ha_1) + mb\,(hb_2 - hb_1) = 0$$

$$mb\,(hb_2 - hb_1) = -\,ma\,(ha_2 - ha_1)$$

$$mb\,(hb_2 - hb_1) = +\,ma\,(ha_1 - ha_2)$$

$$\frac{mb}{ma} = \frac{(ha_1 - ha_2)}{(hb_2 - hb_1)}$$

For $\frac{mb}{ma}$ to be **(+)** then $ha_1 > ha_2$ and $hb_2 > hb_1$

Assume equilibrum $ha_2 = hb_2 = h_2$

$$\frac{mb}{ma} = \frac{(ha_1 - h_2)}{(h_2 - hb_1)}$$

For $\frac{mb}{ma}$ to be **(+)** then $ha_1 > h_2$ and $h_2 > hb_1$

Therefore

$$\boxed{ha_1 > hb_1}$$

2. $hwa_1 > hwb_1$
specific enthalpy work of the source("a") is greater than the specific enthalpy work of the sink ("b")

$Wa + Wb \leq 0$ (from the 2nd law)

$Wa = (Hwa_2 - Hwa_1) = ma (hwa_2 - hwa_1) =$ **(-)**

$Wb = (Hwb_2 - Hwb_1) = mb (hwb_2 - hwb_1) =$ **(+)**

Wa = (-) because media "a" is the source of the energy

$$ma (hwa_2 - hwa_1) + mb (hwb_2 - hwb_1) \leq 0$$

$$ma (hwa_2 - hwa_1) \leq - mb (hwb_2 - hwb_1)$$

$$ma (hwa_2 - hwa_1) \leq + mb (hwb_1 - hwb_2)$$

$$\frac{ma}{mb} \leq \frac{(hwb_1 - hwb_2)}{(hwa_2 - hwa_1)}$$

For $\frac{ma}{mb}$ to be **(+)** and $Wa = ma (hwa_2 - hwa_1) =$ **(-)**

then $hwa_1 > hwa_2$ and $hwb_2 > hwb_1$

Assume equilibrum $hwa_2 = hwb_2 = hw_2$

$$\frac{ma}{mb} \leq \frac{(hwb_1 - hw_2)}{(hw_2 - hwa_1)}$$

For $\frac{ma}{mb}$ to be **(+)** then $hwa_1 > hw_2$ and $hw_2 > hwb_1$

Therefore **$hwa_1 > hwb_1$**

3. $hqa_1 > hqb_1$
specific enthalpy heat of the source("a") is greater than the specific enthalpy heat of the sink ("b")

$Qa + Qb \geq 0$ (from the 2nd law)

$Qa = (Hqa_2 - Hqa_1) = ma (hqa_2 - hqa_1) =$ **(-)**

$qb = (Hqb_2 - Hqb_1) = mb (hqb_2 - hqb_1) =$ **(+)**

Qb = (+) because media "b" is the sink of the energy

$$ma (hqa_2 - hqa_1) + mb (hqb_2 - hqb_1) \geq 0$$

$$mb (hqb_2 - hqb_1) \geq - ma (hqa_2 - hqa_1)$$

$$mb (hqb_2 - hqb_1) \geq + ma (hqa_1 - hqa_2)$$

$$\frac{(hqb_2 - hqb_1)}{(hqa_1 - hqa_2)} \geq \frac{ma}{mb}$$

For $\frac{ma}{mb}$ to be **(+)** and $Qb = mb (hqb_2 - hqb_1) =$ **(+)**

then $hqa_1 > hqa_2$ and $hqb_2 > hqb_1$

Assume equilibrum $hqa_2 = hqb_2 = hq_2$

$$\frac{(hqb_2 - hqb_1)}{(hqa_1 - hqa_2)} \geq \frac{ma}{mb}$$

For $\frac{ma}{mb}$ to be **(+)** then $hqa_1 > hq_2$ and $hq_2 > hqb_1$

Therefore $\boxed{hqa_1 > hqb_1}$

4. $Tmha_1 > Tmhb_1$
mean enthalpy temperature of the source("a") is greater than the mean enthalpy temperature of the sink ("b")

$$\Delta Ha + \Delta Hb = 0 \quad \text{from the energy equation}$$

Remember that: $h = Tmh\,(s)$

The specific enthalpy is equal to the mean enthalpy temperature times the specific entropy.

$\Delta Ha = ma\,(ha_2 - ha_1)$ $\Delta Hb = mb\,(hb_2 - hb_1)$

$ha_1 = Tmha_1\,(sa_1)$ $hb_1 = Tmhb_1\,(sb_1)$

$ha_2 = Tmha_2\,(sa_2)$ $hb_2 = Tmhb_2\,(sb_2)$

$$\Delta Ha + \Delta Hb = 0$$

$$ma\,(Tmha_2\,(sa_2) - Tmha_1\,(sa_1)) + mb\,(Tmhb_2\,(sb_2) - Tmhb_1\,(sb_1)) = 0$$

Assume at equilibrum that

$h_2 = ha_2 = hb_2$ and $Tmh_2 = Tmha_2 = Tmhb_2$ and $s_2 = sa_2 = sb_2$

$$ma\,(Tmh_2\,(s_2) - Tmha_1\,(sa_1)) + mb\,(Tmh_2\,(s_2) - Tmhb_1\,(sb_1)) = 0$$

<u>Add</u> $ma\,(Tmh_2\,(sa_1) - Tmh_2\,(sa_1))$

<u>Add</u> $mb\,(Tmh_2\,(sb_1) - Tmh_2\,(sb_1))$

Notice: both equations are equal to zero so there is no effect on the original equation

Combine the three equations :

$$ma\,(Tmh_2\,(s_2) - Tmha_1\,(sa_1)) + mb\,(Tmh_2\,(s_2) - Tmhb_1\,(sb_1)) +$$
$$ma\,(Tmh_2\,(sa_1) - Tmh_2\,(sa_1)) + mb\,(Tmh_2\,(sb_1) - Tmh_2\,(sb_1)) = 0$$

Combine terms :

$$ma\,(Tmh_2\,(sa_1) - Tmha_1\,(sa_1)) + mb\,(Tmh_2\,(sb_1) - Tmhb_1\,(sb_1)) +$$
$$ma\,(Tmh_2\,(s_2) - Tmh_2\,(sa_1)) + mb\,(Tmh_2\,(s_2) - Tmh_2\,(sb_1)) = 0$$

continue on page 6

Rearranging the equation

$ma\,(sa_1\ (Tmh_2 - Tmha_1)\ ) + mb\,(sb_1\ (Tmh_2 - Tmhb_1)\ )$

$\Rightarrow \quad + ma\ (Tmh_2\,(s_2 - sa_1) + mb\ (Tmh_2\,(s_2 - sb_1) = 0$

$Wa + Wb \leq 0$ and $Qa + Qb \geq 0$ from second law

$Qa + Qb \geq 0$

$Qa = ma\ (hq_2 - hqa_1)$

$Qb = mb\ (hq_2 - hqb_1)$

$ma\ (hq_2 - hqa_1) + mb\ (hq_2 - hqb_1) \geq 0$

Substituting hq with Tr (s)

$ma\ (Tr\,(s_2 - sa_1) + mb\ (Tr\,(s_2 - sb_1) \geq 0$

Multiply each term with $\dfrac{Tmh_2}{Tr}$

$\Rightarrow \quad ma\ (Tmh_2\,(s_2 - sa_1) + mb\ (Tmh_2\,(s_2 - sb_1) \geq 0$

This is equal **part** of the original equation **above**

and if this equation is ≥ 0

Then

$ma\,(sa_1\ (Tmh_2 - Tmha_1)\ ) + mb\,(sb_1\ (Tmh_2 - Tmhb_1)\ ) \leq 0$

With $Wa + Wb \leq 0$

$Wa = ma\,(sa_1\ (Tmh_2 - Tmha_1)\ ) = (-)$ $\quad Tmha_1 > Tmh_2$

$Wb = mb\,(sb_1\ (Tmh_2 - Tmhb_1)\ ) = (+)$ $\quad Tmh_2 > Tmhb_1$

Therefore **$Tmha_1 > Tmhb_1$**

4.5 Weber's 4th Law Thermodynamics

Weber's fourth law ... Equilibrum between two medium

Locating the equilibrum point starts with the energy equation

$$\Delta Ha + \Delta Hb = 0$$

$\Delta Ha = ma\,(ha_2 - ha_1) = (-)$

$\Delta Hb = mb\,(hb_2 - hb_1) = (+)$

$ma\,(ha_2 - ha_1) + mb\,(hb_2 - hb_1) = 0$

Assume: ha_1 and hb_1 are known and at equilibrum that $ha_2 = hb_2 = h_2$

$ma\,(h_2 - ha_1) + mb\,(h_2 - hb_1) = 0$

$h_2\,(ma + mb) = ma\,(ha_1) + mb\,(hb_1)$

$$h_2 = \frac{ma\,(ha_1) + mb\,(hb_1)}{(ma + mb)}$$ final specific enthalpy of source and sink

From the second law

$$\boxed{Qa_1 + Qb_1 \geq 0}$$

$ma\,(hqa_2 - hqa_1) + mb\,(hqb_2 - hqb_1) \geq 0$

Assume at equilibrum that $hqa_2 = hqb_2 = hq_2$

$ma\,(hq_2 - hqa_1) + mb\,(hq_2 - hqb_1) \geq 0$

$hq_2\,(ma + mb) \geq ma\,(hqa_1) + mb\,(hqb_1)$

$$hq_2 \geq \frac{ma\,(hqa_1) + mb\,(hqb_1)}{(ma + mb)}$$

Divide both sides by... Tr ($s = hq / Tr$)

Therefore $$s_2 \geq \frac{ma\,(sa_1) + mb\,(sb_1)}{(ma + mb)}$$

final specific entropy range of source and sink

Locate equilibrum on steam table with two data points h_2 and s_2

4.6 Example of Weber's Laws of Thermodynamics

To understand these new concepts of work and heat lets go back to the example of two containers of water. Below is the construction of a system diagram showing the process of the two mediums and the energy, work and heat transferred.

Source... medium "a" ...mass 2 lbm /hr @ 580 °R

Sink... medium "b" ...mass of 2 lbm/hr @ 540 °R

H2O	Temp	Pressure	Volume	Entropy	Enthalpy	uq / hq	hw	Pv work	Tmh
491.69	°R	psia	ft³/lbm	Btu/lbm-°R	Btu/lbm	Btu/lbm	Btu/lbm	Btu/lbm	°R
1- Initial	580	14.7	0.016	0.16533	88.425	81.29	7.13	0.04	534.84
2- Final	560	14.7	0.016	0.13027	68.445	64.05	4.39	0.04	525.41
(2 - 1)	-20.0	0.0	0.00	-0.03506	-19.980	-17.24	-2.74	0.00	-9.4
2	lbm /hr		0.000	-0.07012	-39.960	-34.48	-5.48	0.00	

Energy	Heat	Work
0.000	1.245	-1.245

H2O	Temp	Pressure	Volume	Entropy	Enthalpy	uq / hq	hw	Pv work	Tmh
491.69	°R	psia	ft³/lbm	Btu/lbm-°R	Btu/lbm	Btu/lbm	Btu/lbm	Btu/lbm	°R
1- Initial	540	14.7	0.016	0.09394	48.465	46.19	2.27	0.04	515.89
2- Final	560	14.7	0.016	0.13027	68.445	64.05	4.39	0.04	525.41
(2 - 1)	20.0	0.0	0.00	0.03633	19.980	17.86	2.12	0.00	9.5
2	lbm /hr		0.000	0.07265	39.960	35.72	4.24	0.00	

Observations from the system diagram:

Two mediums, two process diagrams and two sets of equations that are linked together

Additional information on each process diagram.

Both mediums are "H2O" as displayed with a reference temperature "491.69 °R"

The medium mass "2 lbm / hr" represents the mass of 2 lbm and a time period of one hour.

One of the first requirements of a system is to determine if this is a spontaneous interaction between the mediums. Use the Weber's 3rd Law of thermodynamics to check that the prerequisite for a *thermal* self acting system between medium "a" as the source and medium "b" as the sink have been met.

1. ha1 > hb1 (88.425 > 48.465) ***check***

Specific enthalpy of the source ("a") is greater than the specific enthalpy of sink ("b")

2. hwa1 > hwb1 (7.13 > 2.27) ***check***

Specific enthalpy work of source ("a") is greater than the specific enthalpy work of sink ("b")

3. hqa1 > hqb1 (81.29 > 46.19) ***check*** **Note: sa1 > sb1 (0.16533 > 0.09394)** ***check***

Specific enthalpy heat of source ("a") is greater than the specific enthalpy heat of sink ("b").

4. **Tmha1 > Tmhb1 (534.84 > 515.89)** ***check***

Mean enthalpy temperature of source ("a") is greater than the mean enthalpy temperature of sink ("b")

The prerequisite for a *thermal* self acting system have been confirmed. Next is to determine the equilibrium state between medium "a" and medium "b" using Weber's 4th law.

Final specific enthalpy and entropy of source and sink

$$h_2 = \frac{ma\,(ha_1\ ) + mb\,(hb_1\ )}{(ma + mb\ )} = \frac{(2)\,(88.425\ ) + (2)(48.465\ )}{(2+2\ )} = 68.445 \text{ Btu / lbm}$$

$$s_2 \geq \frac{ma\,(sa_1\ ) + mb\,(sb_1\ )}{(ma + mb\ )} \geq \frac{(2)\,(\ 0.16533\ ) + (2)(\ 0.09394\ )}{(2+2\ )} \geq 0.13027 \text{ Btu / lbm-R}$$

With s_2 having thousands of solutions that are greater than 0.13027
another property must be known... both medium are at a constant pressure of 14.7 psia

Below is a list of possible solutions for h_2 and s_2

	Temp °R	Pressure psia	Volume ft³/lbm	Entropy Btu/lbm-°R	Enthalpy Btu/lbm	uq / hq Btu/lbm	hw Btu/lbm	Pv work Btu/lbm	Tmh °R
	559.68	134.26	0.016	0.12964	68.445	63.74	4.70	0.40	527.96
	559.72	122.45	0.016	0.12970	68.445	63.77	4.67	0.37	527.72
	559.77	103.71	0.016	0.12980	68.445	63.82	4.62	0.31	527.31
	559.81	84.963	0.016	0.12990	68.445	63.87	4.57	0.25	526.91
	559.86	66.217	0.016	0.13000	68.445	63.92	4.53	0.20	526.50
	559.91	47.471	0.016	0.13010	68.445	63.97	4.48	0.14	526.10
	559.96	28.725	0.016	0.13020	68.445	64.02	4.43	0.09	525.69
⇨	560	14.7	0.016	0.13027	68.445	64.05	4.39	0.04	525.41
	560.01	9.9779	0.016	0.13030	68.445	64.07	4.38	0.03	525.29

At a temperature of 560 R @ 14.7 psia the solution is found for h_2 and s_2

Determire if the 1st and 2nd laws are meet.

1. Weber's 1st law of thermodynamics ("Energy can't be created or destroyed")
2. Weber's 2nd law of thermodynamics ("Work can create heat... but heat can not create Work" and "The interaction of two mediums "can not" increase the available work of the system"

1. Weber's 1st law of thermodynamics

Summary : $\Delta Ha + \Delta Hb = 0$

Medium "a"

$\Delta Ha = ma\,(ha_2 - ha_1) = (-)$

$\Delta Ha = (2)\,(68.445 - 88.425) = (-39.960)$

Medium "b"

$\Delta Hb = mb\,(hb_2 - hb_1) = (+)$

$\Delta Hb = (2)\,(68.445 - 48.465) = (+39.960)$

Energy is removed from medium "a" ⇨ Energy is added to medium "b"

Summary : $\Delta Ha + \Delta Hb = 0$

$(-39.960) + (+39.960) = 0$

Did the system meet **Weber's 1st Law of Thermodynamics** ? *check*

2. Weber's 2nd law of thermodynamics

Summary: $Wa + Wb \le 0$ and $Qa + Qb \ge 0$ with $Wa + Qa + Wb + Qb = 0$

Where: Wa + Wb = net work of the system   Qa + Qb= net heat of the system

$Wa + Wb \le 0$

then $ma(hwa_2 - hwa_1) + mb(hwb_2 - hwb_1) \le 0$ (must be (-) or equal to zero)

$Wa = ma (hwa_2 - hwa_1)$ $Wb = mb (hwb_2 - hwb_1)$

$Wa = (2) (4.39 - 7.13) = - 5.48$ $Wb = (2) (4.39 - 2.27) = + 4.24$

$Wa + Wb \le 0$ $(-5.48) + (+4.24) \le 0$ $(- 1.24) \le 0$

$Qa + Qb \ge 0$

then $ma(hqa_2 - hqa_1) + mb(hqb_2 - hqb_1) \ge 0$ (must be (+) or equal to zero)

$Qa = ma (hqa_2 - hqa_1)$ $Qb = mb (hqb_2 - hqb_1)$

$Qa = (2) (64.05 - 81.29) = - 34.48$ $Qb = (2) (64.05 - 46.19) = + 35.72$

$Qa + Qb \ge 0$ $(-34.48) + (+35.72) \ge 0$ $(+ 1.24) \ge 0$

$Wa + Qa + Wb + Qb = 0$

$Wa = (2) (4.39 - 7.13) = - 5.48$

$Qa = (2) (64.05 - 81.29) = - 34.48$

$Wb = (2) (4.39 - 2.27) = + 4.24$

$Qb = (2) (64.05 - 46.19) = + 35.72$

$Wa + Qa + Wb + Qb = (- 5.48) + (- 34.48) + (+4.24) + (+35.72) = 0$

$\triangle Ha + \triangle Hb = Wa + Qa + Wb + Qb = (- 39.960) + (+ 39.960) = 0$

Did the system meet **Weber's 2nd Law of Thermodynamics** ? *check*

Once the system data and calculations meet the Weber's laws of thermodynamics it is required that they are maintained from initial contact to equilibrium. Below are the calculations of medium "a" reducing energy and medium "b" gaining energy at an interval of 3.996 Btu.

Medium "a" 2 lbm /hr @ 580 R

Temp	Pressure	Entropy	Enthalpy	hq /uq	hw
(°R)	(psia)	Btu/lbm-°R	Btu/lbm	(Btu/lbm)	(Btu/lbm)
580.00	14.70	0.16533	88.425	81.29	7.13
578.00	14.70	0.16188	86.427	79.59	6.83
576.00	14.70	0.15842	84.429	77.89	6.54
574.00	14.70	0.15494	82.431	76.18	6.25
572.00	14.70	0.15146	80.433	74.47	5.96
570.00	14.70	0.14796	78.435	72.75	5.68
568.00	14.70	0.14445	76.437	71.02	5.41
566.00	14.70	0.14092	74.439	69.29	5.15
564.00	14.70	0.13739	72.441	67.55	4.89
562.00	14.70	0.13384	70.443	65.81	4.64
560.00	14.70	0.13027	68.445	64.05	4.39

Source intial reading

Source

Energy	Heat	Work
Btu	Btu	Btu
-3.996	-3.393	-0.603
-7.992	-6.795	-1.197
-11.988	-10.217	-1.771
-15.984	-13.639	-2.345
-19.980	-17.081	-2.899
-23.976	-20.533	-3.443
-27.972	-24.004	-3.968
-31.968	-27.476	-4.492
-35.964	-30.967	-4.997
-39.960	-34.477	-5.483

medium "a" first interval calculations

Energy out = (86.427 - 88.425) x 2.0 = - 3.996 Btu

Heat out = (79.59 - 81.29) x 2.0 = - 3.393 Btu

Work out = (6.83 - 7.13) x 2.0 = - 0.603 Btu

Medium "b" 2 lbm /hr @ 540 R

Temp	Pressure	Entropy	Enthalpy	hq /uq	hw
(°R)	(psia)	Btu/lbm-°R	Btu/lbm	(Btu/lbm)	(Btu/lbm)
540.00	14.70	0.09394	48.465	46.19	2.27
542.00	14.70	0.09764	50.463	48.01	2.46
544.00	14.70	0.10132	52.461	49.82	2.64
546.00	14.70	0.10498	54.459	51.62	2.84
548.00	14.70	0.10863	56.457	53.41	3.04
550.00	14.70	0.11227	58.455	55.20	3.25
552.00	14.70	0.11590	60.453	56.99	3.47
554.00	14.70	0.11951	62.451	58.76	3.69
556.00	14.70	0.12311	64.449	60.53	3.92
558.00	14.70	0.12670	66.447	62.30	4.15
560.00	14.70	0.13027	68.445	64.05	4.39

Sink intial reading

Sink

Energy	Heat	Work
Btu	Btu	Btu
3.996	3.632	0.364
7.992	7.253	0.739
11.988	10.853	1.135
15.984	14.442	1.542
19.980	18.021	1.959
23.976	21.591	2.385
27.972	25.141	2.831
31.968	28.681	3.287
35.964	32.212	3.752
39.960	35.722	4.238

medium "b" first interval calculations

Energy in = (50.463 - 48.465) x 2.0 = + 3.996 Btu

Heat in = (48.01 - 46.19) x 2.0 = + 3.632 Btu

Work in = (2.46 - 2.27) x 2.0 = + 0.364 Btu

Below are the "system" calculations of showing the net change in energy, work and heat at an energy interval of 3.996 Btu.

Source Medium "a" 2 lbm /hr @ 580 R				System Interval Changes			**Sink** Medium "b" 2 lbm /hr @ 540 R			
Temp (°R)	Energy Btu	Heat Btu	Work Btu	Energy Btu	Heat Btu	Work Btu	Temp (°R)	Energy Btu	Heat Btu	Work Btu
580.00							540.00			
578.00	-3.996	-3.393	-0.603	0.000	0.239	-0.239	542.00	3.996	3.632	0.364
576.00	-7.992	-6.795	-1.197	0.000	0.458	-0.458	544.00	7.992	7.253	0.739
574.00	-11.988	-10.217	-1.771	0.000	0.635	-0.635	546.00	11.988	10.853	1.135
572.00	-15.984	-13.639	-2.345	0.000	0.802	-0.802	548.00	15.984	14.442	1.542
570.00	-19.980	-17.081	-2.899	0.000	0.940	-0.940	550.00	19.980	18.021	1.959
568.00	-23.976	-20.533	-3.443	0.000	1.058	-1.058	552.00	23.976	21.591	2.385
566.00	-27.972	-24.004	-3.968	0.000	1.137	-1.137	554.00	27.972	25.141	2.831
564.00	-31.968	-27.476	-4.492	0.000	1.206	-1.206	556.00	31.968	28.681	3.287
562.00	-35.964	-30.967	-4.997	0.000	1.245	-1.245	558.00	35.964	32.212	3.752
560.00	-39.960	-34.477	-5.483	0.000	1.245	-1.245	560.00	39.960	35.722	4.238

"System" first interval calculations

Net Energy = ΔHa + ΔHb = (-3.996 + 3.996) = 0 Btu

Net Work = Wa + Wb = Δ Hwa + Δ Hwb = (- 0.603 + 0.364) = - 0.239 Btu

Net Heat = Qa + Qb = ΔHqa + Δ Hqb = (-3.393 + 3.632) = + 0.239 Btu

Spontaneous interaction between two mediums "a" and "b":

1. Energy moves from a higher level "a" to a lower level "b" (ha1 >hb1)

2. Specific work energy of "a" is greater than medium "b" (hwa1 >hwb1)

3. Specific heat energy of "a" is greater than medium "b" (hqa1 >hqb1)

4. Mean enthalpy temperature of "a" is greater than medium "b" (Tmha1 >Tmhb1)

5. Change in energy of "a" is plus energy of medium "b" is equal to zero

6 Work of "a" plus Work of "b" is less than or equal to zero ($Wa + Wb \leq 0$)

7. Heat of "a" plus Heat of "b" is greater than or equal to zero ($Qa + Qb \geq 0$)

8. Sum of the work and heat is equal to zero… ($Wa + Wb + Qa + Qb = 0$)

9. Items 1 thru 8 were maintained from initial contact to equilibrium. *check*

5 Work and Heat Applications

"Thermodynamics isn't harder than rocket science, it is rocket science"
Fred J. Weber

5.1 Applying Work and Heat

The following section is applying the work and heat energy to different applications such as a heat and work exchanger. Also in many applications of thermodynamics there is a close relationship between thermal energy and mechanical energy in such items as nozzle, turbine, pump and compressor. The following is general energy with thermal and mechanical energy:

A system follows the general energy equation.

$\Delta Ea + \Delta Eb = 0$ general energy equation

General energy equation represents all energy but the focus is the relationship between thermal energy and mechanical energy

Where: ME (mechanical energy) = KE (kinetic energy) + PE = potential energy

Change in Mechical energy is equal to

$$\Delta ME = \Delta KE + \Delta PE$$

Combining the thermal and mechanical energy to the general equation

$$\Delta Ea + \Delta Eb = 0$$

Source "a" $\Delta Ea = \Delta TEa + \Delta KEa + \Delta PEa$

Sink "b" $\Delta Eb = \Delta TEb + \Delta KEb + \Delta PEb$

$$\Delta TEa + \Delta KEa + \Delta PEa + \Delta TEb + \Delta KEb + \Delta PEb = 0$$

Work = Change in Work Energy involved

$$\Delta TEwa + \Delta KEwa + \Delta PEwa + \Delta TEwb + \Delta KEwb + \Delta PEwb = (-)$$

Heat = Change in Heat Energy involved

$$\Delta TEqa + \Delta KEqa + \Delta PEqa + \Delta TEqb + \Delta KEqb + \Delta PEqb = (+)$$

Change in Energy = change in Work energy + the change in Heat energy = Work + Heat

$$\Delta E = \Delta Ew + \Delta Eq = W + Q$$

Rules for transferring energy between a source and sink

1. Energy moves from a higher level to a lower level
2. Total change in energy is zero
3. Sum of the work and heat is equal to zero... $Wa + Qa + Wb + Qb = 0$
4. Net work is (-) or zero...work removed by the system
5. Net heat is (+) or zero...heat added by the system
6. Summary...Net work is (-) + Net heat is (+) is equal to zero

Below are the calculations for kinetic energy and potential energy. Keep in mind that ME (mechanical energy) = KE (kinetic energy) + PE (potential energy)

Kinetic Energy

KE (kinetic energy)= (mass) (specific kinetic energy) = (m) (ke)

$$\Delta KE = (m)(\Delta ke)$$

change in specifc kinetic energy = $(\Delta ke) = \dfrac{\mathbf{v}^2}{2\, g_c\, J}$

Let $\mathbf{v}$ = velocity (ft / sec) $g_c = \dfrac{32.2 \text{ lbm - ft}}{\text{lbf - sec}^2}$ = gravitational constant

$J = \dfrac{778 \text{ ft - lbf}}{\text{Btu}}$ = Joule's constant (ft-lbf to Btu)

$$\Delta ke \text{ (Btu / lbm)} = \frac{\mathbf{v}^2 \quad (\text{ft}^2 / \text{sec}^2)}{(g)\dfrac{32.2 \text{ lbm - ft}}{\text{lbf - sec}^2} \quad (J)\dfrac{778 \text{ ft - lbf}}{\text{Btu}}} = \frac{\mathbf{v}^2}{(25051.6)}$$

Potential Energy

PE (potential energy)= (mass) (specific potential energy) = (m) (pe)

$$\Delta PE = (m)(\Delta pe)$$

change in specifc potential energy = $(\Delta ke) = \dfrac{\mathbf{h}\, g}{J\ g_c}$

Let $\mathbf{h}$ = height (ft) $g_c = \dfrac{32.2 \text{ lbm - ft}}{\text{lbf - sec}^2}$ = gravitational constant

$J = \dfrac{778 \text{ ft - lbf}}{\text{Btu}}$ = Joule's constant (ft-lbf to Btu) $g = \dfrac{32.2 \text{ ft}}{\text{sec}^2}$ = acceleration

$$\Delta pe \text{ (Btu / lbm)} = \frac{(g)\dfrac{32.2 \text{ ft}}{\text{sec}^2} \quad \mathbf{h} \text{ (ft)}}{(g_c)\dfrac{32.2 \text{ lbm - ft}}{\text{lbf - sec}^2} \quad (J)\dfrac{778 \text{ ft - lbf}}{\text{Btu}}} = \frac{\mathbf{h}}{(778)}$$

With the addition of different types of energy there is a need for a process diagram that shows these changes. Below is a process diagram with thermal and mechanical energy (kinetic and potential).

Thermo-diagram with TE, KE and PE

	Thermal Energy					Kinetic Energy				Potential Energy			
H2O	Temp	Pressure	Enthalpy	uq / hq	hw	Velocity	ket	keq	kew	height	pet	peq	pew
491.69	°R	psia	Btu/lbm	Btu/lbm	Btu/lbm	ft/sec	Btu/lbm	Btu/lbm	Btu/lbm	ft	Btu/lbm	Btu/lbm	Btu/lbm
1- Initial	1121.80	2400.00	719.15	444.15	275.00	10.00	0.00	0.00	0.00	5.00	0.00	0.00	0.00
2- Final	2716.20	2400.00	2243.60	944.39	1299.21	50.00	0.10	0.00	0.10	100.00	0.13	0.00	0.13
(2 - 1)	1594.40	0.00	1524.45	500.24	1024.21	40.00	0.10	0.00	0.10	95.00	0.13	0.00	0.13
20	lbm /hr		30489	10005	20484		2	0	2		3	0	3

The example above shows energy added to the medium at a constant pressure with an increase in velocity and elevation.

This approach maybe used to describe what happens when a ball is dropped, hits the floor and doesn't bounce back to the same elevation. The ball starts at a maximum PE.. While falling the ball's PE is converted to KE. The ball hits the floor at a maximum KE. The KE is converted to TE but some of the TE work energy is converted to TE heat energy due to the floor. The TE converts back to KE and PE but the ball doesn't reach the same elevation.

Before moving into different applications of heat and work energy obtain a Temperature and specific entropy graph of water and steam. Follow each point on the graph.

5.2 Heat and Work Exchanger

The "heat and work exchanger" is the replacement of the "heat exchanger". The new name is a reminder that heat and work are both transferred, not just heat. An example of this could be the two containers of water from the previous section. A symbol is used to show the flow of energy in and out of the exchanger. The symbol for a heat and work exchanger is:

Heat and Work Exchanger

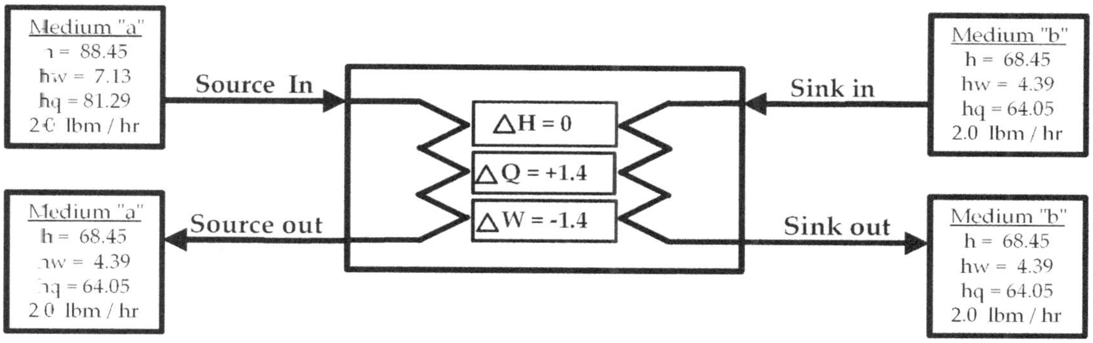

The heat and work exchanger is the equipment used in industry to exchanger heat and work between mediums. Two common types of exchangers are the plate and the shell and tube. With a heat and work exchangers keep in mind...

Spontaneous interaction between two mediums "a" and "b":

1. Energy moves from a higher level "a" to a lower level "b" (ha1 >hb1)
2. Specific work energy of "a" is greater than medium "b" (hwa1 >hwb1)
3. Specific heat energy of "a" is greater than medium "b" (hqa1 >hqb1)
4. Mean enthalpy temperature of "a" is greater than medium "b" (Tmha1 >Tmhb1)
5. Change in energy of "a" is plus energy of medium "b" is equal to zero
6 Work of "a" plus Work of "b" is less than or equal to zero ($Wa + Wb \leq 0$)
7. Heat of "a" plus Heat of "b" is greater than or equal to zero ($Qa + Qb \geq 0$)
8. Sum of the work and heat is equal to zero... ($Wa + Wb + Qa + Qb = 0$)
9. Items 1 thru 8 were maintained from initial contact to equilibrium

Example of a heat and work exchanger.

Medium "a".... the Source

H2O	Temp	Pressure	Volume	Entropy	Enthalpy	uq / hq	hw	Pv work	Tmh
491.69	°R	psia	ft³/lbm	Btu/lbm-°R	Btu/lbm	Btu/lbm	Btu/lbm	Btu/lbm	°R
1- Initial	825.45	150	3.053	1.5768	1200.00	775.30	424.70	84.72	761.04
1	lbm /hr								

Medium "b".... the Sink

H2O	Temp	Pressure	Volume	Entropy	Enthalpy	uq / hq	hw	Pv work	Tmh
491.69	°R	psia	ft³/lbm	Btu/lbm-°R	Btu/lbm	Btu/lbm	Btu/lbm	Btu/lbm	°R
1- Initial	576.57	14.7	0.016	0.1594	85.00	78.38	6.62	0.04	533.22
10	lbm /hr								

Spontaneous interaction between two mediums "a" and "b":

1. Energy moves from a higher level "a" to a lower level "b" (ha1 >hb1) (1200.00 > 85.00) Btu/lbm
2. Specific work energy of "a" is greater than medium "b" (hwa1 >hwb1) (424.70 > 6.62) Btu/lbm
3. Specific heat energy of "a" is greater than medium "b" (hqa1 >hqb1) (775.30 > 78.38) Btu/lbm
4. Mean enthalpy temperature of "a" is greater than medium "b" (Tmha1 >Tmhb1) (761.04 > 533.22) R

Using Weber's fourth law, the equilibrum between two mediums and solve for h_2 and s_2 range

$$h_2 = \frac{ma\,(ha_1) + mb\,(hb_1)}{(ma + mb)}$$ final specific enthalpy of source and sink

$$h_2 = \frac{1\,(1200.00) + 10\,(85.00)}{(1 + 10)} = 186.36 \text{ Btu / lbm}$$ $h_2 = ha_2 = hb_2$

$$s_2 \geq \frac{ma\,(sa_1) + mb\,(sb_1)}{(ma + mb)}$$ final specific entropy range of source and sink

$$s_2 \geq \frac{1\,(1.5768.) + 10\,(0.1594)}{(1 + 10)} = 0.28826 \text{ Btu / lbm - R}$$ min value of sa_2 and sb_2

Locate equilibrum on steam table with two data points h and P

Locate equilibrum points of medium "a" using ha_2 and the pressure of 150 psia

Locate equilibrum points of medium "b" using hb_2 and the pressure of 14.7 psia

Check that $sa_2 > s_2$ and $sb_2 > s_2$

The heat and work exchanger is represented by the "system diagram" shown by the interaction of two constant pressure processes.

H2O	Temp	Pressure	Volume	Entropy	Enthalpy	uq / hq	hw	Pv work	Tmh
491.69	°R	psia	ft³/lbm	Btu/lbm-°R	Btu/lbm	Btu/lbm	Btu/lbm	Btu/lbm	°R
1- Initial	825.45	150	3.053	1.5768	1200.00	775.30	424.70	84.72	761.04
2- Final	677.36	150	0.017	0.3208	186.36	157.71	28.65	0.46	581.00
(2 - 1)	-148.1	0.0	-3.04	-1.2560	-1013.64	-617.58	-396.06	-84.26	-180.0
1	lbm /hr		-3.036	-1.2560	-1013.64	-617.58	-396.06	-84.26	

Energy	Heat	Work
0.0	179.0	-179.0

H2O	Temp	Pressure	Volume	Entropy	Enthalpy	uq / hq	hw	Pv work	Tmh
491.69	°R	psia	ft³/lbm	Btu/lbm-°R	Btu/lbm	Btu/lbm	Btu/lbm	Btu/lbm	°R
1- Initial	576.57	14.7	0.016	0.1594	85.00	78.38	6.62	0.04	533.22
2- Final	671.64	14.7	0.184	0.3214	186.36	158.04	28.32	0.50	579.80
(2 - 1)	95.1	0.0	0.17	0.1620	101.36	79.66	21.70	0.46	46.6
10	lbm /hr		1.679	1.6201	1013.60	796.59	217.01	4.57	

Display both processes in equal intervals of "change in Enthalpy"

Note: the source is intervals will be 10x that of the sink

Equilibrium at ↔ The source h, hq, hw and Tmh is greater than the sink

Temp	Entropy	Enthalpy	hq /uq	hw	Tmh		Temp	Entropy	Enthalpy	hq /uq	hw	Tmh
(°R)	Btu/lbm-R	Btu/lbm	Btu/lbm	Btu/lbm	°R		(°R)	Btu/lbm-R	Btu/lbm	Btu/lbm	Btu/lbm	°R
825.45	1.5768	1200.00	775.30	424.70	761.04		576.57	0.1594	85.00	78.38	6.62	533.22
818.09	1.3290	997.27	653.46	343.81	750.39		596.85	0.1940	105.27	95.37	9.90	542.74
818.09	1.0812	794.54	531.62	262.92	734.87		617.12	0.2274	125.55	111.80	13.75	552.18
818.09	0.8334	591.82	409.79	182.03	710.10		637.34	0.2597	145.82	127.69	18.13	561.52
818.09	0.5856	389.09	287.94	101.15	664.41		657.52	0.2910	166.09	143.08	23.01	570.76
678.35	0.3222	187.36	158.44	28.92	581.45	↔	671.64	0.3213	186.26	157.97	28.29	579.76
677.36	0.3208	186.36	157.71	28.65	581.01		671.64	0.3214	186.36	158.03	28.33	579.82

Notice the equilibrium point doesn't reach h @ 186.36 Btu lbm as shown in the system diagram. The hw value of the source needs to be greater than the hw of the sink.

The heat and work exchanger can operate in two different flow patterns. With a co-current flow exchanger the source and sink flows are in the same direction across the exchanger. The co-current mode is seldom used in industry. The counter-current flow exchanger, the source and sink flows are in the opposite direction across the exchanger. Notice the change in work and heat for these two different flow patterns of operation.

Heat and Work Exchanger @ Co-Current Flow

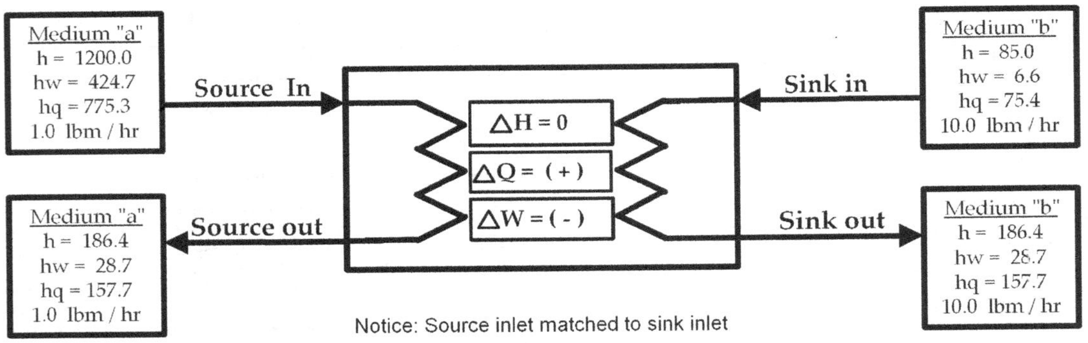

Notice: Source inlet matched to sink inlet

Medium "a" @ 150 psia							Medium "b" @ 14.7 psia			
Temp (°R)	Enthalpy Btu/lbm	hq /uq (Btu/lbm)	hw (Btu/lbm)	Energy (Btu)	Heat (Btu)	Work (Btu)	Temp (°R)	Enthalpy (Btu/lbm)	hq /uq (Btu/lbm)	hw (Btu/lbm)
825.45	1200.000	775.30	424.70				576.57	85.000	78.38	6.62
818.09	997.270	653.46	343.81				596.85	105.270	95.37	9.90
-7.36	-202.73	-121.84	-80.89				20.28	20.27	16.99	3.28
	-202.73	-121.84	-80.89	0.0	48.0	-48.1		202.70	169.88	32.82
818.09	997.270	653.46	343.81				596.85	105.270	95.37	9.90
818.09	794.540	531.62	262.92				617.11	125.550	111.80	13.75
0.00	-202.73	-121.84	-80.89				20.26	20.28	16.43	3.85
	-202.73	-121.84	-80.89	0.1	42.4	-42.4		202.80	164.27	38.53
818.09	794.540	531.62	262.92				617.11	125.550	111.80	13.75
818.09	591.820	409.78	182.04				637.34	145.820	127.69	18.13
0.00	-202.72	-121.83	-80.89				20.23	20.27	15.89	4.38
	-202.72	-121.83	-80.89	0.0	37.1	-37.1		202.70	158.91	43.79
818.09	591.820	409.78	182.04				637.34	145.820	127.69	18.13
818.09	389.090	287.94	101.15				657.52	166.090	143.09	23.00
0.00	-202.73	-121.84	-80.89				20.18	20.27	15.40	4.87
	-202.73	-121.84	-80.89	0.0	32.2	-32.2		202.70	154.00	48.70
818.09	389.090	287.94	101.15				657.52	166.090	143.09	23.00
677.36	186.360	157.71	28.65				671.64	186.360	158.04	28.32
-140.73	-202.73	-130.23	-72.50				14.12	20.27	14.95	5.32
	-202.73	-130.23	-72.50	0.0	19.3	-19.3		202.70	149.52	53.18

Below is data for a counter-current flow heat and work exchanger. Notice the net work and heat differences compared to the co-current flow exchanger.

Heat and Work Exchanger @ Counter-Current Flow

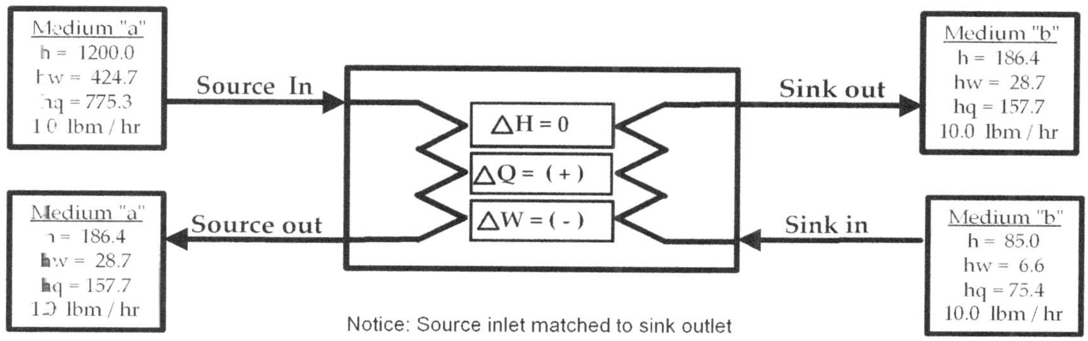

Notice: Source inlet matched to sink outlet

Medium "a" @ 150 psia							Medium "b" @ 14.7 psia			
Temp	Enthalpy	hq /uq	hw	Energy	Heat	Work	Temp	Enthalpy	hq /uq	hw
(°R)	Btu/lbm	Btu/lbm	Btu/lbm	Btu	Btu	Btu	(°R)	Btu/lbm	Btu/lbm	Btu/lbm
825.45	1200.000	775.30	424.70				657.52	166.090	143.09	23.00
818.09	997.270	653.46	343.81				671.64	186.360	158.04	28.32
-7.36	-202.73	-121.84	-80.89				14.12	20.27	14.95	5.32
	-202.73	-121.84	-80.89	0.0	27.7	-27.7		202.70	149.52	53.18
818.09	997.270	653.46	343.81				637.34	145.820	127.69	18.13
818.09	794.540	531.62	262.92				657.52	166.090	143.09	23.00
0.00	-202.73	-121.84	-80.89				20.18	20.27	15.40	4.87
	-202.73	-121.84	-80.89	0.0	32.2	-32.2		202.70	154.00	48.70
818.09	794.540	531.62	262.92				617.11	125.550	111.80	13.75
818.09	591.820	409.78	182.04				637.34	145.820	127.69	18.13
0.00	-202.72	-121.83	-80.89				20.23	20.27	15.89	4.38
	-202.72	-121.83	-80.89	0.0	37.1	-37.1		202.70	158.91	43.79
818.09	591.820	409.78	182.04				596.85	105.270	95.37	9.90
818.09	389.090	287.94	101.15				617.11	125.550	111.80	13.75
0.00	-202.73	-121.84	-80.89				20.26	20.28	16.43	3.85
	-202.73	-121.84	-80.89	0.0	42.4	-42.4		202.80	164.27	38.53
818.09	389.090	287.94	101.15				576.57	85.000	78.38	6.62
677.36	186.360	157.71	28.65				596.85	105.270	95.37	9.90
-140.73	-202.73	-130.23	-72.50				20.28	20.27	16.99	3.28
	-202.73	-130.23	-72.50	0.0	39.6	-39.7		202.70	169.88	32.82

Below is another example of a heat exchanger with the low pressure side at a higher energy level than the high pressure side. Notice at an ideal equilibrium point of h= 186.36 Btu / lbm the sink hw value (28.65) is greater than the source hw value of 28.32 Btu / lbm. Therefore the new equilibrium points: source at h = 189.36 Btu / lbm and sink at h= 186.06 Btu / lbm.

H2O	Temp	Pressure	Volume	Entropy	Enthalpy	uq / hq	hw	Pv work	Tmh
491.69	°R	psia	ft³/lbm	Btu/lbm-°R	Btu/lbm	Btu/lbm	Btu/lbm	Btu/lbm	°R
1- Initial	773.31	14.7	31.090	1.8256	1200.00	897.63	302.37	84.55	657.32
2- Final	671.64	14.7	0.184	0.3214	186.36	158.04	28.32	0.50	579.80
(2 - 1)	-101.7	0.0	-30.91	-1.5042	-1013.64	-739.59	-274.05	-84.05	-77.5
1	lbm /hr		-30.906	-1.5042	-1013.64	-739.59	-274.05	-84.05	

Energy	Heat	Work
0.0	57.2	-57.3

H2O	Temp	Pressure	Volume	Entropy	Enthalpy	uq / hq	hw	Pv work	Tmh
491.69	°R	psia	ft³/lbm	Btu/lbm-°R	Btu/lbm	Btu/lbm	Btu/lbm	Btu/lbm	°R
1- Initial	576.22	150	0.016	0.1587	85.00	78.03	6.97	0.45	535.60
2- Final	677.36	150	0.017	0.3208	186.36	157.71	28.65	0.46	581.00
(2 - 1)	101.1	0.0	0.00	0.1621	101.36	79.68	21.68	0.02	45.4
10	lbm /hr		0.006	1.6206	1013.60	796.83	216.77	0.16	

Display both processes in equal intervals of "change in Enthalpy"
Note: the source is intervals will be 10x that of the sink
Equilibrium at ↔ The source h, hq, hw and Tmh is greater than the sink

Temp	Entropy	Enthalpy	hq /uq	hw	Tmh		Temp	Entropy	Enthalpy	hq /uq	hw	Tmh
(°R)	Btu/lbm-R	Btu/lbm	Btu/lbm	Btu/lbm	°R		(°R)	Btu/lbm-R	Btu/lbm	Btu/lbm	Btu/lbm	°R
773.31	1.8256	1200.00	897.63	302.37	657.32		576.22	0.1587	85.00	78.03	6.97	535.60
671.64	1.5288	997.27	751.70	245.57	652.32		596.51	0.1933	105.27	95.03	10.24	544.65
671.64	1.2269	794.54	603.25	191.29	647.60		616.79	0.2267	125.55	111.47	14.08	553.79
671.64	0.9251	591.82	454.86	136.96	639.74		637.02	0.2591	145.82	127.37	18.45	562.90
671.64	0.6233	389.09	306.45	82.64	624.29		657.22	0.2904	166.09	142.77	23.32	571.99
671.64	0.3274	190.36	160.96	29.40	581.48		676.96	0.3202	185.96	157.42	28.54	580.83
671.64	0.3259	189.36	160.23	29.13	581.07	↔	677.06	0.3203	186.06	157.49	28.57	580.87
671.64	0.3244	188.36	159.50	28.86	580.66		677.16	0.3205	186.16	157.56	28.60	580.93
671.64	0.3229	187.36	158.77	28.59	580.24		677.26	0.3206	186.26	157.64	28.62	580.97
671.64	0.3214	186.36	158.04	28.32	579.80		677.36	0.3208	186.36	157.71	28.65	581.00

System diagram with the new equilibrium points ↔

H2O	Temp	Pressure	Volume	Entropy	Enthalpy	uq / hq	hw	Pv work	Tmh
491.69	°R	psia	ft³/lbm	Btu/lbm-°R	Btu/lbm	Btu/lbm	Btu/lbm	Btu/lbm	°R
1- Initial	773.31	14.7	31.090	1.8256	1200.00	897.63	302.37	84.55	657.32
2- Final	671.64	14.7	0.267	0.3259	189.36	160.23	29.13	0.73	581.07
(2 - 1)	-101.7	0.0	-30.82	-1.4997	-1010.64	-737.40	-273.24	-83.82	-76.2
1	lbm /hr		-30.823	-1.4997	-1010.64	-737.40	-273.24	-83.82	

Energy	Heat	Work
0.0	57.2	-57.3

H2O	Temp	Pressure	Volume	Entropy	Enthalpy	uq / hq	hw	Pv work	Tmh
491.69	°R	psia	ft³/lbm	Btu/lbm-°R	Btu/lbm	Btu/lbm	Btu/lbm	Btu/lbm	°R
1- Initial	576.22	150	0.016	0.1587	85.00	78.03	6.97	0.45	535.60
2- Final	677.06	150	0.017	0.3203	186.06	157.49	28.57	0.46	580.87
(2 - 1)	100.8	0.0	0.00	0.1616	101.06	79.46	21.60	0.02	45.3
10	lbm /hr		0.006	1.6161	1010.60	794.62	215.98	0.16	

5.3 Nozzle and Orifice

The nozzle and orifice are simply mechanical devices that change the medium's properties with no outside energy source and sink. In the discussion of "system" there was an emphasis on the source and sink but the nozzle and orifice are an application where the system is self contained that is when no energy leaves the device.

The orifice destroys energy by converting the thermal work energy into thermal heat energy. The orifice is a constant enthalpy process where the change in enthalpy work is equal to the change in enthalpy heat. The orifice can also be called the "zero efficient nozzle". A symbol for the orifice is shown below:

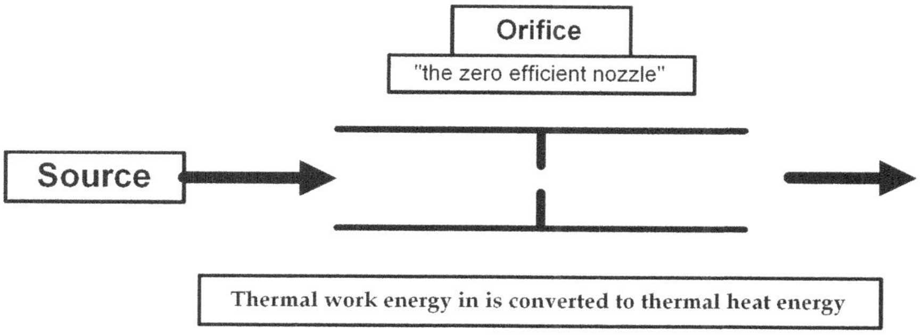

Below is a system diagram with an orifice. The change in enthalpy is zero and there is no effect from a source or sink. Note the change in kinetic energy is zero. The change in enthalpy work (-88.5 Btu / lbm) is convert to a change in enthalpy heat (+88.5 Btu / lbm).

Orifice

	Thermal energy								Kinetic energy				
H2O	Temp	Pressure	Volume	Entropy	Enthalpy	uq / hq	hw	Tmh	Velocity	ket	keq	kew	Flow
491.69	°R	psia	ft³/lbm	Btu/lbm-°R	Btu/lbm	Btu/lbm	Btu/lbm	°R	ft/sec	Btu/lbm	Btu/lbm	Btu/lbm	Btu/hr
1- Initial	1398.5	100.00	8.29	1.90000	1501.70	934.21	567.49	790.37	0.00	0.00	0.00	0.00	
2- Fina	1393.6	19.48	42.58	2.08000	1501.70	1022.72	478.98	721.97	0.00	0.00	0.00	0.00	0.00
(2 - 1)	-4.9	-80.53	34.29	0.18000	0.00	88.50	-88.50	-68.40	0.00	0.00	0.00	0.00	0.00

thermal work energy is converted to thermal heat energy

Change in Enthalpy is zero...
no thermal energy added to or removed from medium

Change in velocity is zero...
no kinetic energy added to or removed from medium

The nozzle converts thermal energy to kinetic energy to be used for a turbine. The change in thermal energy available for work is equal to the gain in kinetic energy available to perform work. Below is a symbol for the nozzle.

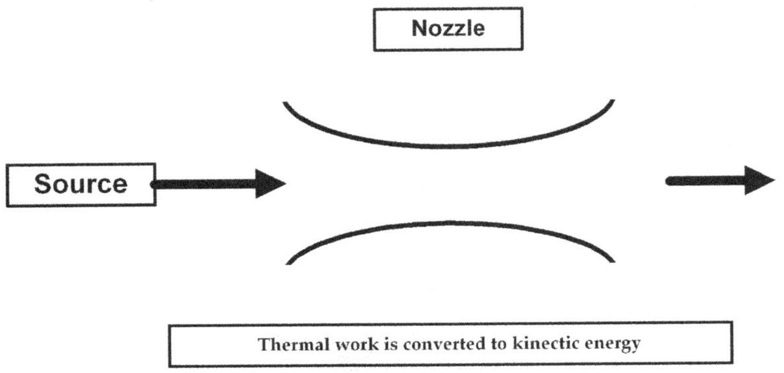

The ideal nozzle operates as a constant entropy process. Below is a system diagram of a 100 percent efficient nozzle. There is no effect from a source or sink. The change in enthalpy (-88.5 Btu / lbm) is converted to the change in kinetic energy (+88.5 Btu / lbm) with a net change in energy of zero.

Nozzle

	Thermal energy								Kinetic energy				
H2O	Temp	Pressure	Volume	Entropy	Enthalpy	uq / hq	hw	Tmh	Velocity	ket	keq	kew	Flow
491.69	°R	psia	ft³/lbm	Btu/lbm-°R	Btu/lbm	Btu/lbm	Btu/lbm	°R	ft/sec	Btu/lbm	Btu/lbm	Btu/lbm	Btu/hr
1- Initial	1398.5	100.00	8.29	1.90000	1501.70	934.21	567.49	790.37	0.00	0.00	0.00	0.00	0
2- Final	1220.9	54.00	13.40	1.90000	1413.20	934.21	478.99	743.79	2105.39	88.50	0.00	88.50	12327
(2 - 1)	-177.6	-46.00	5.12	0.00000	-88.50	0.00	-88.50	-46.58	2105.39	88.50	0.00	88.50	12327

thermal energy removed from medium is converted to kinetic energy

No energy leaves the medium

Below are the equations that support the nozzle design.

Nozzle Equations

$\Delta Ea + \Delta Eb = 0$ general energy equation

$\Delta Eb = 0$ no source or sink acting on medium ...self contained

$\Delta Ea = \Delta TEa + \Delta KEa = 0$

$$\Delta Ea = ma\,(ha_2 - ha_1) + ma\left(\frac{v^2 a_2}{2gJ} - \frac{v^2 a_1}{2gJ}\right) = 0$$

Let v = velocity (ft / sec)

$g = \dfrac{32.2 \text{ lbm - ft}}{\text{lbf - sec}^2}$ = gravitational constant

$J = \dfrac{778 \text{ ft - lbf}}{\text{Btu}}$ = Joule's constant (ft-lbf to Btu)

$$ma\,(ha_2 - ha_1) + ma\left(\frac{v^2 a_2}{2gJ} - \frac{v^2 a_1}{2gJ}\right) = 0$$

Divide both sides by ma and Assume inlet velocity small $va_1 = 0$

$$(ha_2 - ha_1) + \left(\frac{v^2 a_2}{2gJ}\right) = 0$$

$$v^2 a_2 = (2gJ)(ha_1 - ha_2)$$

$$va_2 = \sqrt{(2gJ)(ha_1 - ha_2)}$$

$$va_2 = (223.8)\sqrt{(ha_1 - ha_2)} \quad (\text{ft / sec})$$

<u>Continuity equation</u>... medium flow through a nozzle

$$F = \left(\frac{(Aa_2)(v\ a_2)(3600)}{(va_2)}\right) \quad (\text{lbm / hr})$$

Let F = flow (lbm / hr)

va_2 = final specific volume (ft^3 / lbm)

Aa_2 = nozzle area (ft^2)

(3600 sec / 1 hr)

v = velocity (ft / sec)

One of the first steps in the design of a nozzle is to locate the maximum flow point. With a known inlet source the maximum flow was determine to be between 0.5 to 0.6 pressure ratio based on St. Venant's equation. The pressure ratio is outlet pressure/ inlet pressure and from the calculations below the maximum flow occurred at 54 psia / 100 psia equal to a 0.54 pressure ratio.

.

Nozzle peak flow selection

Medium: Steam Size : 2 inch dia.

	Thermal energy								Kinetic energy				
H2O	Temp	Pressure	Volume	Entropy	Enthalpy	uq / hq	hw	Tmh	Velocity	ket	keq	kew	Flow
491.69	°R	psia	ft³/lbm	Btu/lbm-°R	Btu/lbm	Btu/lbm	Btu/lbm	°R	ft/sec	Btu/lbm	Btu/lbm	Btu/lbm	Btu/hr
1- Initial	1398.5	100.00	8.288	1.9000	1501.70	934.21	567.49	790.37	0.00	0	0.00	0.00	
2- Final	1249.9	60.00	12.350	1.9000	1427.50	934.21	493.29	751.32	1927.80	74.20	0.00	74.20	12251
2- Final	1240.5	58.00	12.680	1.9000	1422.90	934.21	488.69	748.89	1986.66	78.80	0.00	78.80	12296
2- Final	1230.8	56.00	13.031	1.9000	1418.10	934.21	483.89	746.37	2046.27	83.60	0.00	83.60	12324
2- Final	1220.9	54.00	13.404	1.9000	1413.20	934.21	478.99	743.79	2105.39	88.50	0.00	88.50	12327
2- Final	1210.6	52.00	13.803	1.9000	1408.20	934.21	473.99	741.16	2164.04	93.50	0.00	93.50	12304
2- Final	1200	50.00	14.230	1.9000	1403.00	934.21	468.79	738.42	2223.41	98.70	0.00	98.70	12262

$$\mathbf{v}a_2 = (223.8)\sqrt{(ha_1 - ha_2)} \quad (\text{ft / sec})$$

$$\mathbf{F} = \left(\frac{(Aa_2)(\mathbf{v}\ a_2)(3600)}{(va_2)}\right) \quad (\text{lbm / hr})$$

Maximum flow

First calculation

$$\mathbf{v}a_2 = (223.8)\sqrt{(1501.7 - 1427.5)} = (223.8)(8.614) = 1927.8 \ (\text{ft / sec})$$

$$\mathbf{F} = \frac{(0.0218)(1927.8)(3600)}{(12.350)} = 12250.5 \ (\text{lbm / hr})$$

$(Aa_2) = (r^2)(3.14) = 3.14\ \text{in}^2 = 0.0218\ \text{ft}^2$ **Nozzle size : 2 inch dia.**

The maximum flow point 12327 lbm /hr is located at 54 psia with a velocity of 2105.39 ft / sec.

Next is to determine the efficiency of a nozzle.

Nozzle efficiency

Nozzle design
Medium: Steam
Size : 2 inch dia.

100 % efficiency

					Thermal energy					Kinetic energy			
H2O	Temp	Pressure	Volume	Entropy	Enthalpy	uq / hq	hw	Tmh	Velocity	ket	keq	kew	Flow
491.69	°R	psia	ft³/lbm	Btu/lbm-°R	Btu/lbm	Btu/lbm	Btu/lbm	°R	ft/sec	Btu/lbm	Btu/lbm	Btu/lbm	Btu/hr
1- Initial	1398.5	100.00	8.29	1.90000	1501.70	934.21	567.49	790.37	0.00	0.00	0.00	0.00	0
2- Final	1220.9	54.00	13.40	1.90000	1413.20	934.21	478.99	743.79	2105.39	88.50	0.00	88.50	12327
(2 - 1)	-177.6	-46.00	5.12	0.00000	-88.50	0.00	-88.50	-46.58	2105.39	88.50	0.00	88.50	12327

75 % efficiency

H2O	Temp	Pressure	Volume	Entropy	Enthalpy	uq / hq	hw	Tmh	Velocity	ket	keq	kew	Flow
491.69	°R	psia	ft³/lbm	Btu/lbm-°R	Btu/lbm	Btu/lbm	Btu/lbm	°R	ft/sec	Btu/lbm	Btu/lbm	Btu/lbm	Btu/hr
1- Initial	1398.5	100.00	8.29	1.90000	1501.70	934.21	567.49	790.37	0.00	0.00	0.00	0.00	0
2- Final	1264.2	42.19	17.79	1.94500	1435.50	956.34	479.16	738.05	1820.91	66.20	0.00	66.20	8031
(2 - 1)	-134.3	-57.31	9.51	0.04500	-66.20	22.13	-88.33	-52.32	1820.91	66.20	0.00	66.20	8031

50 % efficiency

H2O	Temp	Pressure	Volume	Entropy	Enthalpy	uq / hq	hw	Tmh	Velocity	ket	keq	kew	Flow
491.69	°R	psia	ft³/lbm	Btu/lbm-°R	Btu/lbm	Btu/lbm	Btu/lbm	°R	ft/sec	Btu/lbm	Btu/lbm	Btu/lbm	Btu/hr
1- Initial	1398.5	100.00	8.29	1.90000	1501.70	934.21	567.49	790.37	0.00	0.00	0.00	0.00	0
2- Final	1307.7	32.80	23.70	1.99000	1457.70	978.46	479.24	732.51	1484.52	44.00	0.00	44.00	4916
(2 - 1)	-90.8	-67.20	15.41	0.09000	-44.00	44.25	-88.25	-57.86	1484.52	44.00	0.00	44.00	4916

25 % efficiency

H2O	Temp	Pressure	Volume	Entropy	Enthalpy	uq / hq	hw	Tmh	Velocity	ket	keq	kew	Flow
491.69	°R	psia	ft³/lbm	Btu/lbm-°R	Btu/lbm	Btu/lbm	Btu/lbm	°R	ft/sec	Btu/lbm	Btu/lbm	Btu/lbm	Btu/hr
1- Initial	1398.5	100.00	8.29	1.90000	1501.70	934.21	567.49	790.37	0.00	0.00	0.00	0.00	0
2- Final	1351.2	25.37	31.68	2.03500	1480.00	1000.59	479.41	727.27	1042.53	21.70	0.00	21.70	2583
(2 - 1)	-47.3	-74.63	23.39	0.13500	-21.70	66.38	-88.08	-63.10	1042.53	21.70	0.00	21.70	2583

0 % efficiency

H2O	Temp	Pressure	Volume	Entropy	Enthalpy	uq / hq	hw	Tmh	Velocity	ket	keq	kew	Flow
491.69	°R	psia	ft³/lbm	Btu/lbm-°R	Btu/lbm	Btu/lbm	Btu/lbm	°R	ft/sec	Btu/lbm	Btu/lbm	Btu/lbm	Btu/hr
1- Initial	1398.5	100.00	8.29	1.90000	1501.70	934.21	567.49	790.37	0.00	0.00	0.00	0.00	0
2- Final	1393.6	19.48	42.58	2.08000	1501.70	1022.72	478.98	721.97	0.00	0.00	0.00	0.00	0
(2 - 1)	-4.9	-80.53	34.29	0.18000	0.00	88.50	-88.50	-68.40	0.00	0.00	0.00	0.00	0

The nozzle is designed for a maximum efficiency at a change in enthalpy work of (– 88.5 Btu /lbm) at a final hw value of 479 Btu/ lbm. The efficiency curve is a constant hw curve. As the nozzle efficiency drops the change in enthalpy work of (– 88.5 Btu /lbm) is maintained and enthalpy heat is added. The enthalpy heat is added until the change in enthalpy work (– 88.5 Btu /lbm) and the change in enthalpy heat is (+ 88.5 Btu /lbm) at this point the nozzle is an orifice.

5.4 Pump and Compressor

The pump and compressor are devices that take mechanical energy and convert it to thermal energy with the focus to increase the enthalpy work of the medium. The most efficient way to increase the enthalpy work of a medium for any mechanical to thermal conversion is by a constant entropy process. Any process outside the constant entropy process adds enthalpy heat and reduces the effectiveness of the energy added to the medium. A symbol for a pump and compressor is shown below:

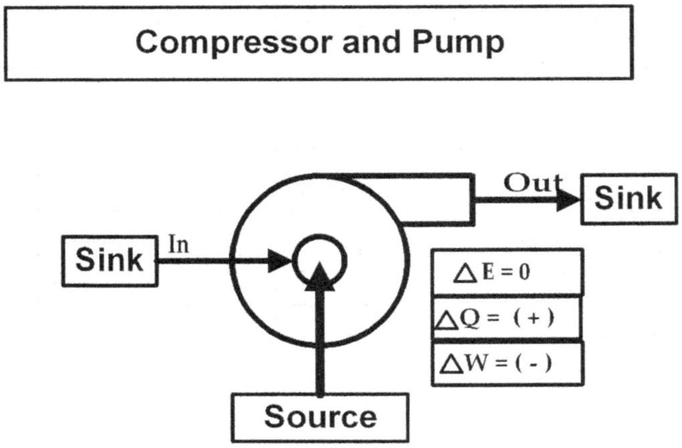

The centrifugal pumps and compressors are widely used mechanical devices used to add energy. Their design is based on taking rotation kinetic energy and transferring the energy to the medium to be used later in the cycle. Without getting into the design of these devices the next pages show the effect of the medium at various efficiencies when a constant energy source is applied. Notice the change in enthalpy is constant due to a constant load being applied. As the efficiency drops the enthalpy work transferred to the medium drops due to the increase in enthalpy heat.

Below is an example of a compressor's efficiency. Notice the change in enthalpy is constant 144.40 Btu / lbm which is the energy supplied by the source (motor /compressor).

Compressor efficiency

Compressor design
Medium: Steam
5 to 1 compression ratio
Input 144.4 Btu /lbm

100 % efficiency — Thermal energy

H2O	Temp	Pressure	Volume	Entropy	Enthalpy	uq / hq	hw	Tmh
491.69	°R	psia	ft³/lbm	Btu/lbm-°R	Btu/lbm	Btu/lbm	Btu/lbm	°R
1- Initial	677.96	15.00	26.52	1.76000	1154.10	865.37	288.73	655.74
2- Final	991.85	75.00	7.76	1.76000	1298.50	865.37	433.13	737.78
(2 - 1)	313.9	60.00	-18.76	0.00000	144.40	0.00	144.40	82.05

75 % efficiency

H2O	Temp	Pressure	Volume	Entropy	Enthalpy	uq / hq	hw	Tmh
491.69	°R	psia	ft³/lbm	Btu/lbm-°R	Btu/lbm	Btu/lbm	Btu/lbm	°R
1- Initial	677.96	15.00	26.52	1.76000	1154.10	865.37	288.73	655.74
2- Final	985.4	38.26	15.22	1.83340	1298.50	901.46	397.04	708.25
(2 - 1)	307.4	23.26	-11.29	0.07340	144.40	36.09	108.31	52.51

50 % efficiency

H2O	Temp	Pressure	Volume	Entropy	Enthalpy	uq / hq	hw	Tmh
491.69	°R	psia	ft³/lbm	Btu/lbm-°R	Btu/lbm	Btu/lbm	Btu/lbm	°R
1- Initial	677.96	15.00	26.52	1.76000	1154.10	865.37	288.73	655.74
2- Final	982.1	19.59	29.74	1.90680	1298.50	937.55	360.95	680.98
(2 - 1)	304.1	4.59	3.22	0.14680	144.40	72.18	72.22	25.25

25 % efficiency

H2O	Temp	Pressure	Volume	Entropy	Enthalpy	uq / hq	hw	Tmh
491.69	°R	psia	ft³/lbm	Btu/lbm-°R	Btu/lbm	Btu/lbm	Btu/lbm	°R
1- Initial	677.96	15.00	26.52	1.76000	1154.10	865.37	288.73	655.74
2- Final	980.39	10.05	57.99	1.98030	1298.50	973.69	324.81	655.71
(2 - 1)	302.4	-4.95	31.47	0.22030	144.40	108.32	36.08	-0.03

0 % efficiency

H2O	Temp	Pressure	Volume	Entropy	Enthalpy	uq / hq	hw	Tmh
491.69	°R	psia	ft³/lbm	Btu/lbm-°R	Btu/lbm	Btu/lbm	Btu/lbm	°R
1- Initial	677.96	15.00	26.52	1.76000	1154.10	865.37	288.73	655.74
2- Final	979.52	5.16	112.95	2.05370	1298.50	1009.78	288.72	632.27
(2 - 1)	301.6	-9.84	86.43	0.29370	144.40	144.41	-0.01	-23.47

Notice at 50 percent efficiency that half the 144.4 Btu / lbm applied to medium is converted into enthalpy work. The other half is converted to enthalpy heat.

Below is an example of a pump's efficiency. Notice the change in enthalpy is constant 4.15 Btu / lbm due to the energy supplied by the source (motor / pump).

Pump efficiency

Pumpdesign
Medium: Water
Input 4.15 Btu /lbm

Thermal energy

100 % efficiency

H2O	Temp	Pressure	Volume	Entropy	Enthalpy	uq / hq	hw	Tmh
491.69	°R	psia	ft³/lbm	Btu/lbm-°R	Btu/lbm	Btu/lbm	Btu/lbm	°R
1- Initial	516.46	100.00	0.02	0.04935	25.17	24.26	0.90	509.97
2- Final	516.63	1500.00	0.02	0.04935	29.31	24.26	5.05	593.97
(2 - 1)	0.2	1400.00	0.00	0.00000	4.15	0.00	4.15	84.00

75 % efficiency

H2O	Temp	Pressure	Volume	Entropy	Enthalpy	uq / hq	hw	Tmh
491.69	°R	psia	ft³/lbm	Btu/lbm-°R	Btu/lbm	Btu/lbm	Btu/lbm	°R
1- Initial	516.46	100.00	0.02	0.04935	25.17	24.26	0.90	509.97
2- Final	517.68	1131.30	0.02	0.05146	29.31	25.30	4.01	569.64
(2 - 1)	1.2	1031.30	0.00	0.00211	4.15	1.04	3.11	59.67

50 % efficiency

H2O	Temp	Pressure	Volume	Entropy	Enthalpy	uq / hq	hw	Tmh
491.69	°R	psia	ft³/lbm	Btu/lbm-°R	Btu/lbm	Btu/lbm	Btu/lbm	°R
1- Initial	516.46	100.00	0.02	0.04935	25.17	24.26	0.90	509.97
2- Final	518.73	762.23	0.02	0.05356	29.31	26.34	2.97	547.21
(2 - 1)	2.3	662.23	0.00	0.00422	4.15	2.07	2.07	37.24

25 % efficiency

H2O	Temp	Pressure	Volume	Entropy	Enthalpy	uq / hq	hw	Tmh
491.69	°R	psia	ft³/lbm	Btu/lbm-°R	Btu/lbm	Btu/lbm	Btu/lbm	°R
1- Initial	516.46	100.00	0.02	0.04935	25.17	24.26	0.90	509.97
2- Final	519.77	392.90	0.02	0.05567	29.31	27.37	1.94	526.49
(2 - 1)	3.3	292.90	0.00	0.00632	4.15	3.11	1.04	16.52

0 % efficiency

H2O	Temp	Pressure	Volume	Entropy	Enthalpy	uq / hq	hw	Tmh
491.69	°R	psia	ft³/lbm	Btu/lbm-°R	Btu/lbm	Btu/lbm	Btu/lbm	°R
1- Initial	516.46	100.00	0.02	0.04935	25.17	24.26	0.90	509.97
2- Final	520.82	23.38	0.02	0.05778	29.31	28.41	0.90	507.30
(2 - 1)	4.4	-76.62	0.00	0.00843	4.15	4.15	0.00	-2.67

At 75 percent efficiency, 75 percent of the energy is converted to enthalpy work.

5.5 Cycle of Systems

In thermodynamics there are discussions about the efficiency of a cycle. With a new understanding of a cycle and system it becomes apparent that the energy sources and sink determine the efficiency of a cycle. Without knowing the energy removed from and added to the processes of a cycle is flawed. It is impossible to determine that one cycle is more efficient than another without knowing the energy consumed by the cycle. Below is a simplified Rankine cycle for a typical power plant with the energy transferred.

Rankine Cycle / Systems

Process 1 to 2

From Process 4 to 1

H2O	Temp	Pressure	Volume	Entropy	Enthalpy	uq / hq	hw	Pv work	Tmh
491.69	°R	psia	ft³/lbm	Btu/lbm-°R	Btu/lbm	Btu/lbm	Btu/lbm	Btu/lbm	°R
1- Initial	585.7	2	0.016	0.1751	94.09	86.09	8.00	0.01	537.35
2- Final	586.8	2400	0.016	0.1751	101.26	86.09	15.17	7.16	578.30
(2 - 1)	1.1	2398.0	0.00	0.0000	7.17	0.00	7.17	7.15	40.9
1	lbm /hr		0.000	0.0000	7.17	0.00	7.17	7.15	

Energy	Heat	Work
0.0	0.0	0.0

Pump

Mech	met	meq	mew
Energy	Btu/lbm	Btu/lbm	Btu/lbm
1- Initial	108.43	0.00	108.43
2- Final	101.26	0.00	101.26
(2 - 1)	-7.17	0.00	-7.17
1	-7.17	0.00	-7.17

Process 2 to 3

H2O	Temp	Pressure	Volume	Entropy	Enthalpy	uq / hq	hw	Pv work	Tmh
491.69	°R	psia	ft³/lbm	Btu/lbm-°R	Btu/lbm	Btu/lbm	Btu/lbm	Btu/lbm	°R
1- Initial	586.8	2400	0.016	0.1751	101.26	86.09	15.17	7.16	578.30
2- Final	1460	2400	0.322	1.5346	1462.80	754.55	708.25	142.96	953.21
(2 - 1)	873.2	0.0	0.31	1.3595	1361.54	668.45	693.09	135.80	374.9
1	lbm /hr		0.306	1.3595	1361.54	668.45	693.09	135.80	

Energy	Heat	Work
0.0	224.4	-224.4

Boiler

Thermo	Enthalpy	uq / hq	hw
Energy	Btu/lbm	Btu/lbm	Btu/lbm
1- Initial	1598.95	798.95	800.00
2- Final	1462.80	754.55	708.25
(2 - 1)	-136.15	-44.40	-91.75
10	-1361.54	-444.03	-917.47

To Process 3 to 4

cont'.....Rankine Cycle / Systems

Process 3 to 4

From Process 2 to 3

H2O	Temp	Pressure	Volume	Entropy	Enthalpy	uq / hq	hw	Pv work	Tmh
491.69	°R	psia	ft³/lbm	Btu/lbm-°R	Btu/lbm	Btu/lbm	Btu/lbm	Btu/lbm	°R
1- Initial	1460	2400	0.322	1.5346	1462.80	754.55	708.25	142.96	953.21
2- Final	585.7	2	135.290	1.5346	890.33	754.55	135.78	50.06	580.17
(2 - 1)	-874.3	-2398.0	134.97	0.0000	-572.47	0.00	-572.47	-92.90	-373.0
1	lbm /hr		134.968	0.0000	-572.47	0.00	-572.47	-92.90	

Energy	Heat	Work
0.0	120.0	-120.0

Turbine

Mech Energy	met Btu/lbm	meq Btu/lbm	mew Btu/lbm
1- Initial	0	0.00	0.00
2- Final	572.47	120.00	452.47
(2 - 1)	572.47	120.00	452.47
1	572.47	120.00	452.47

Process 4 to 1

H2O	Temp	Pressure	Volume	Entropy	Enthalpy	uq / hq	hw	Pv work	Tmh
491.69	°R	psia	ft³/lbm	Btu/lbm-°R	Btu/lbm	Btu/lbm	Btu/lbm	Btu/lbm	°R
1- Initial	585.7	2	135.290	1.5346	890.33	754.55	135.78	50.06	580.17
2- Final	585.7	2	0.016	0.1751	94.09	86.09	8.00	0.01	537.35
(2 - 1)	0.0	0.0	-135.27	-1.3595	-796.24	-668.45	-127.79	-50.05	-42.8
1	lbm /hr		-135.274	-1.3595	-796.24	-668.45	-127.79	-50.05	

Energy	Heat	Work
0.0	15.2	-15.2

Condenser

H2O	Temp	Pressure	Volume	Entropy	Enthalpy	uq / hq	hw	Pv work	Tmh
491.69	°R	psia	ft³/lbm	Btu/lbm-°R	Btu/lbm	Btu/lbm	Btu/lbm	Btu/lbm	°R
1- Initial	560	14.7	0.016	0.1303	68.44	64.05	4.39	0.04	525.40
2- Final	585.7	14.7	0.016	0.1751	94.12	86.10	8.02	0.04	537.50
(2 - 1)	25.7	0.0	0.00	0.0448	25.68	22.05	3.63	0.00	N / A
31.01	lbm /hr		0.003	1.3905	796.27	683.69	112.59	0.01	

Back To Process 1 to 2

As for a reverse-Rankine cycle, the energy source and sink are a compressor, two heat and work exchangers, and an orifice. This cycle is used for refrigeration and HVAC.

A question…
How can a cycle with two inefficient constant temperature processes, a sink that absorbs an unlimited amount of energy and a source that supplies an unlimited amount of energy be the most efficient cycle? Or is the Carnot cycle is limiting our thinking.

5.6 Final Note and Thank you

Final Note

This book covers a significant amount of information about a slightly different approach to Thermodynamics. I will continue to update the book as I go though my notes. Because we just touch the surface there is more to cover. Therefore, future editions of this book will cover:

1. Comments and correction of this book
2. New phases such as power work, power heat, and enthalpy temperature.
3. Different medium …Air, Ammonia ($NH3$), R134 and etc
4. Mixture mediums…Psychometrics (Air and Water), Absorption (Ammonia and Water)
5. Applications of different processes, cycles and systems (power plant, refrigeration and etc.)
6. Other applications of work and heat energy

Thank you

Thank you for support so we can continue supplying hands-on information to you. Next, I hope with this information we were able to take care of your priorities:

"Take care of yourself and your family"… hopefully the book made life easier for you and gives you more time to spend with your family.

"Take of your co-worker"… that you used something in this book to make life better for your engineering and maintenance team.

"Operate the plant meeting safety and environmental requirements while achieving a profit"… implementing information from this book increased the safety and environmental awareness in your plant, while improving the overall plant performance.

If nothing else, the book gets you thinking about thermodynamics. To quote Eleanor Roosevelt again … "Great minds discuss ideas; average minds discuss events; small minds discuss people."

For Young engineers only…

One of the most important decisions a young engineer can make coming out of college is to determine what company to choose to start your career. Don't make a job decision based on money, perks, or location. Instead, select the job based on the *manager* you will be working for because this person will be your mentor and prepare you the best for the future.

For managers only…

"Delegating …giving a person the responsibility, knowledge and opportunity to do something but supplying a net to catch them if they need it!"

Good luck in your career. Remember to take a break occasionally, grab a cup of coffee, and get one for me. Take care of yourself and your family… Fred

Index

Made in the USA
Monee, IL
08 November 2026

67971104R00083